The Age of Electroacoustics

Transformations: Studies in the History of Science and Technology

Jed Z. Buchwald, general editor

For a complete list of books in the series, please see the back of the book.

The Age of Electroacoustics

Transforming Science and Sound

Roland Wittje

The MIT Press
Cambridge, Massachusetts
London, England

This book was set in Stone Sans and Stone Serif by Toppan Best-set Premedia Limited. Printed and bound in the United States of America.

Library of Congress Cataloging-in-Publication Data is available.

ISBN: 978-0-262-03526-2

10 9 8 7 6 5 4 3 2 1

Contents

Acknowledgments

Many have contributed to this book. Unfortunately, it is nearly impossible to mention every individual, and I apologize to those whom I have left out. This book project began more than ten years ago with a postdoctoral position at the Forum for the History of Knowledge at the Norwegian University of Science and Technology (NTNU). I would like to thank especially Anders Johnsson and the exhibition team of *Etterklang: Vitenskap, musikk og massemedia i elektroakustikkens tidsalder* (Reverberations: Science, music and mass media in the age of electroacoustics), an exhibition we put up at NTNU Vitenskapsmuseet in 2007.

I started to formulate my ideas for the book, especially chapters 4 and 5, in my manuscript for the workshop "Acoustics, Applied and Pure" at the Dibner Institute in Cambridge, Massachusetts, in 2006. I would like to thank all the participants, as well as David Hounshell, Klaus Hentschel, Michael Eckert, Emily Thompson, and anonymous referees for their comments on the manuscript. Among my numerous involvements in conferences, I enjoyed especially the sound studies sessions of the International Committee for the History of Technology (ICOHTECH), organized by Hans-Joachim Braun. I want to thank Wolfgang König and Friedrich Steinle for two productive research stays at the Technical University of Berlin. During my research as a scholar-in-residence at the Deutsches Museum in Munich in 2007, I received generous support from Oskar Blumtritt.

From October 2007 to April 2014, I taught at the History of Science Unit of the University of Regensburg. I would like to thank Christoph Meinel, who gave me all the support and freedom that I needed; our secretary, Angelika Sonntag; and all my students. During my stay as long-term research scholar at the Science Museum, London, in 2011, I received great support from Jane Wess. Two papers published in edited volumes contain some of the ideas and arguments elaborated in this book. "The Electrical Imagination: Sound Analogies, Equivalent Circuits, and the Rise of Electroacoustics, 1863–1939" has been published in *Osiris*, volume 28, "Music, Sound, and the Laboratory from 1750 to 1980" (Wittje 2013). "Concepts and Significance of Noise in Acoustics: Before and after the Great War" has been published in *Perspectives on Science*, volume 24/1, a special issue on "Understanding Noise in Twentieth-Century Physics and Engineering" (Wittje 2016). I would like to thank the editors of the *Osiris* volume—Alexandra Hui, Myles Jackson, and Julia Kursell—and the organizer of the *Perspectives on Science* issue, Chen-Pang Yeang, as well as the participants of the Max Planck Institute workshop in preparation for the *Osiris* volume for discussions and comments.

From March to May 2015, I was visiting researcher at the Epistemes of Modern Acoustics group at the Max Planck Institute for the History of Science in Berlin, where I had the opportunity to finish this manuscript. I would like to thank Viktoria Tkaczyk, others in the group, and Lino Camprubi for their comments on chapter 3. A big thanks goes to the people who read and commented on the manuscript: Christian Reiss, who has read all of it; Kate Sturge, who worked marvelously on the introduction and chapters 2 to 4; and Thomas Steinhauser, David Pantalony, and Jed Buchwald. I would especially like to thank all the people who have helped me at the archives and collections.

Last but not least, I would like to thank Jahnavi, Neelambari, and Revati for their great support, love, and patience.

Chennai, 23 January 2016
Roland Wittje

List of Figures

1   Introduction: A History and Geography of Acoustics

When I began my research on Johan Holtsmark's acoustics laboratory at the Norwegian Institute of Technology more than fifteen years ago, I did not know what I would get myself into. I wondered, why would an accomplished physicist like Holtsmark engage in the research field of acoustics—apparently outdated by 1930—while building Scandinavia's first particle accelerator? What unfolded was a fascinating story that has occupied me ever since. Although much has been published on the history of early particle accelerators, historians of science have written surprisingly little about acoustics in the twentieth century, especially about Holtsmark's German colleagues, to whom he regularly referred. This is the beginning of the story of the present book.

Just a few decades into the twentieth century, acoustics had changed from a scientific field based on the understanding and performance of classical music into one guided by electrical engineering and media technologies: electroacoustics. *The Age of Electroacoustics* traces these transformations of acoustics from the times of Hermann Helmholtz and Lord Rayleigh, in the 1860s and 1870s, to the end of the interwar period, with the outbreak of World War II.[1] Throughout this history, acoustics crossed into and brought together diverse academic disciplines, from physics to music, aesthetics, physiology, psychology, phonetics, and electrical engineering.

While keeping sight of the multidisciplinary character of acoustics, I concentrate in this book on physical acoustics and on how physicists oriented themselves and framed their study of sound within a larger academic, scientific, and cultural discourse. I aim to excavate the layers of transformation that acoustics underwent—as a set of scientific practices and as a scientific community—in its different institutional settings, in its epistemology, in the kinds of sounds it dealt with, and, not least, in the soundscapes it helped to shape. I try to discover how the political, social, military, cultural, and economic threads of this history have transformed our understanding of acoustics as a science, and how electroacoustic media technologies, acoustic knowledge, and acoustic practices were coproduced.

How does a historian of physics come to write about the history of twentieth-century acoustics? Acoustics is hardly the first subject that comes to mind when thinking about the history of science in the early twentieth century. Several decades into the history of science's turn toward practice and culture, our image of modern physics in the interwar period continues to be guided by a history of ideas rather than of practice.[2] The development of quantum mechanics, relativity, and atomic and nuclear physics still dominates our image of physics since the turn of the twentieth century.

Yet if we stop focusing on a normative understanding of what modern physics should be and start looking at what physicists in the early twentieth century actually did, how they did it, and why, a different picture unfolds.[3] It then emerges that the standard narrative has sidelined a wide range of experimental, applied, and industrial research fields where the majority of physicists worked. This narrow understanding of modern physics in historiography, and the resulting under-representation of certain fields, is partly caused by tensions between what was considered at that time to be basic and applied physics within the physics community, and a hierarchy that places basic and theoretical physics above experimental and applied research.[4]

Just as historians of science have argued that categories of classical versus modern physics are highly problematic in our histories (Staley 2005), similar reservations are in order for the categories of basic and applied research (Shapin 2004). Still, though scholars from science and technology studies, cultural historians, and historians of technology, music, and media have written extensively about the modern soundscape, historians of twentieth-century physics are largely absent in the thriving field of sound studies. Welcome exceptions are Emily Thompson, with her outstanding *The Soundscape of Modernity* (2002), and some acousticians who engaged in the history of their field, such as Frederick V. Hunt and Robert T. Beyer.[5]

It remains to connect these histories of acoustics with a history of twentieth-century physics more broadly. My study of the history of physics at the Norges Tekniske Høgskole (Norwegian Institute of Technology) in the interwar period (Wittje 2003) revealed that research in acoustics and the construction of a particle accelerator for experimental nuclear physics went hand in hand, both relying on a common set of experimental and scientific instrument practices. An intriguing story emerged of acoustics research driven by the advent of radio broadcasting and sound motion pictures. In Norway, most of the impulses for the new acoustics research and for media technologies came from Central Europe and the United States. Although Norway was not a combatant in World War I, in Germany, Britain, and other countries, the employment of acoustic technologies and methods in warfare played a key role in the transformation of acoustics as a discipline. The Norwegian example demonstrated that there was no inherent separation between fields of atomic and nuclear physics and applied fields such as technical acoustics between the wars, and that modern physics was coproduced by a broad spectrum of research fields and practices. Of course, there were many tensions between physicists from the different fields, as well as divergent visions of what modern physics should be. The boundaries of what physics should be, how it should be practiced, and how it actually was practiced were constantly being redrawn.[6]

These observations led to some of the questions that I attempt to answer in this book. What was the place of acoustics within the discipline of physics in the period, and how should the transformations of the field of acoustics be located within the transformations of physics more generally? Did acoustics become a scientific domain of its own? Was it seen as proper, academic physics? What was its relationship with the physics community at large? What effect did acoustics have on our historical understanding of physics at the time? With these questions in mind, the emergence of technical acoustics and electroacoustics offers a case study of a scientific field located between science and engineering in interwar Germany. My study thereby aims to contribute to the historiography of twentieth-century science, creating links to a cultural history of sound and sound studies in other disciplines.

1.1   Sound and Acoustics

What do we mean by sound? What do we mean by acoustics? Sound might be defined as the physical phenomena we perceive through our sense of hearing; acoustics might then be defined as the science of sound. Neither of these notions is self-evident, and neither is unproblematic. What has been perceived and defined as sound, which sounds have been studied, and how they have been perceived and studied have differed considerably over space and time. Musician R. Murray Schafer introduced the concept of the sound-scape as a sonic environment in the 1960s (Schafer 1994; E. Thompson 2002, 2). In the following decades, historians from media, music and cul-tural studies have begun to historicize the modern soundscape and the ways in which people listen.[7] Along with the soundscape and listening practices, there were changes in the epistemic objects of acoustics research and in the knowledge that was created through and about those epistemic objects. Acoustics can be understood as a body of knowledge, as a commu-nity of practitioners, and as a set of practices. Accordingly, a range of differ-ent questions arise: What were the sound objects that became the topic of acoustics research? What were the practices of knowledge generation, who created the knowledge, and for what purposes was it created? What were the cultural, political, economic, and technological circumstances of the generation of sound and the generation of knowledge about sound? In this book, I try to find meaningful answers to these questions and to combine the different perspectives. Investigating acoustics as the scientific engage-ment with sound, I trace the transformation of that scientific interest and its entanglement with the transformation of the modern soundscape in one of acoustics' most vibrant periods: from the late nineteenth century to the end of the interwar years.

The existing history of acoustics shows a gap between the acoustics of the late nineteenth century and acoustics research in the 1920s and 1930s. The transformations between the two, I claim, have not yet been fully understood—whether the industrialization of acoustics, its conversion into electroacoustics, or the role World War I played in these transformations. Until the outbreak of World War I, the notion of acoustics as a field and of the activities of acousticians revolved around the classic works by Herman Helmholtz and Lord Rayleigh, including the questions they raised. Helm-holtz, the "Bismarck of physics" and first president of the Physikalisch-Technische Reichsanstalt (Imperial Institute of Physics and Technology), and Rayleigh, the successor of James Clerk Maxwell as Cavendish Professor of Physics at Cambridge, were leading authorities in the physics communi-ties of Wilhelmine Prussia and Victorian Britain, respectively.

Throughout the nineteenth century, acoustics was a subdiscipline of physics that, because of its intrinsic relation to the human sense of hearing and the theory and practice of music, could not be reduced to physics alone but maintained essential connections to physiology and musicology. With psychophysics and the emergence of experimental psychology in the late nineteenth century, psychoacoustics became another field for the acoustician to occupy.[8] In the 1890s, physicists started to turn away from acoustics as electrical research—especially in electromagnetic wave propagation, thermodynamics, and new entities such as X-rays and radioactivity— became the new frontiers of physics research. The unresolved debates around Helmholtz's concepts of consonance, dissonance, and combination tones were displaced into the realm of the emerging discipline of experimental psychology.

Acousticians themselves have conventionally perceived World War I as the turning point for a revival of acoustics research after decades of stagnation.[9] Yet, as historians we need to ask whether acoustics as a science had really stagnated until then, and if so, what exactly had stagnated? What were the forces and the nature of transformation during the war? Who were the actors of the new acoustics, and what kind of practices did they contribute and develop? How was the transformation of acoustics related to other transformations both inside and outside science in the late nineteenth and early twentieth century? Certainly, the transformation of acoustics was accelerated by scientists' confrontation with new sounds and new demands placed on sound measurement in warfare, but the process of change had started long before the war. The industrialization of acoustics and its transformation into electroacoustics in fact began with the inventions of the phonograph, telephone, and wireless, as well as the rise of the electrical industry. Sound motion pictures and radio, the main transformers of acoustics in the interwar period, were also conceived before the outbreak of the war. Whereas we generally think of the war as an accelerator of technological development, the development of sound motion pictures in Europe was probably slowed rather than accelerated by the war and its consequences.

The crisis of European bourgeois society that questioned the dichotomy of musical sounds and noise had also manifested itself in art and music already before the outbreak of the war. Research interest in acoustics started to grow during and after World War I, when acoustics was heavily employed in war research and the development of war technologies, including submarine detection, air raid surveillance, artillery, and telecommunication.[10] Acoustics finally experienced extraordinary growth and substantial transformation from the mid-1920s onward, mainly because of the breakthrough of media technologies and sound amplification. At the same time, the development of new electroacoustic measurement technologies fundamentally altered the character of acoustics research.

1.2   A Geography of Acoustics: Sound, Space, and Time

The local production of scientific knowledge and technologies, as well as their diffusion, circulation, and exchange through national and transnational regimes and networks of exchange, have become central themes in the history of science and technology (Livingstone 2003; Shapin 1998; Sabra 1996). Although I have not developed the geography of acoustics into an analytical framework for my narrative, the geographical aspect needs to be addressed. The stories of the transformation of acoustics narrated in this book draw on examples and developments mainly from German-speaking research communities, but also from Britain and North America. A geography of acoustics can be understood as a geography of sound—of the geographical spread of sounds, soundscapes, and sound technologies. How did sounds and soundscapes differ across various geographical and cultural regions? How did they travel and circulate? On the other hand, a geography of acoustics can also be approached as a geography of knowledge about sound. This geography of knowledge then becomes both a physical and a mental map. Acoustic knowledge circulated through publications and other means of communication, through the travel of scientists, and through the spread of acoustic technologies. In terms of disciplinary geography, acoustics as a set of knowledges and practices shifted its boundaries within the different disciplines, institutions, and technologies with which it was entangled.

Acoustic knowledge and technologies of mass media, in particular, spread rapidly across political, economic, and cultural boundaries. At the same time, local and national political, economic, and technocratic regimes and interests guided and controlled these flows. Historians of physics have discussed scientific nationalism versus scientific internationalism in Europe before and after World War I (Forman 1973; Carson, Kojevnikov, and Trischler 2011). These histories have focused mainly on the community of theoretical physicists around the development of quantum mechanics and the attitudes of scientists toward scientific universalism on the one hand, growing nationalism and political hostility—right up to boycotts and counter-boycotts—on the other. While many interwar acousticians were also involved in these battles between nationalism, boycott, and internationalism, the spread of acoustic knowledge during and after World War I followed different patterns. The circulation of acoustic knowledge and technologies was not guided by the sentiments and norms of a scientific community separate from society at large. To understand the production, consumption, and circulation of acoustic knowledge and practices during and after World War I, it is necessary to view them as crucial elements of large technological systems of significant economic and military importance.

The concept of large technological systems was introduced by Thomas P. Hughes (1983, 1987), drawing on the example of electric light and power systems:

Technological systems contain messy, complex, problem solving components. They are both socially constructed and socially shaping. Among the components in technological systems are physical artifacts. ... Technological systems also include organizations, such as manufacturing firms, utility companies, and investment banks, and they incorporate components usually labeled scientific, such as books, articles, and university teaching and research programs. (1987, 51)

Economic, state, and military interests can either accelerate or obstruct the flow of knowledge and technologies within and among these systems. Acousticians, acoustic knowledge and practices, and acoustic technologies became part of large technological systems during World War I. New weapons such as aircraft, submarines, and heavy artillery could operate only within these large technological systems. The main systems of warfare that required and transformed acoustic knowledge and practices were artillery ranging for trench warfare, submarine warfare, and air defense. These large technological systems of warfare and their requirements drove the transformation and industrialization of acoustics during World War I.

After the war, the large technological systems of mass media replaced those of warfare as the main driving force of acoustics research. The three main technological systems of mass media that incorporated acoustic knowledge, artifacts, and practices were telephony, radio broadcasting, and sound motion pictures. While the composition and dynamics of large technological systems were obviously different during wartime and peace, many similarities and continuities led from the war into the peacetime economy. During World War I, when sound location was used for detecting the enemy, acoustic knowledge could be lifesaving or life threatening: detecting an enemy submarine could save the lives of sailors and passengers, but the same acoustic knowledge threatened the lives of the crew of the submarine to be hunted down. Patents were kept secret, sensitive research was not published, and taking manuals to the forward trenches was forbidden. Nevertheless, the military observed the activities of the enemy, and scientists and engineers on all sides were generally quick to figure out and assess the enemy's methods and technologies. The various parties discussed and developed the same or very similar types of technologies, as David Bloor has pointed out for the case of aircraft location (Bloor 2000). Measures of acoustic location and their countermeasures were, effectively, coproduced.

Metaphors of war and peace were widely used in international economic battles before and after the war, as in the Marconi patent war and the sound motion picture wars that concluded with the Paris Sound-Film Peace Treaty of 1930 (Mühl-Benninghaus 1999; Distelmeyer 2003; Hong 2001). To be well informed about the latest international developments in acoustics and able to compete internationally in acoustics research and development was, therefore, also a matter of life and death for the motion picture and electrical industries, even if not in the same sense as for the soldiers and civilian victims of World War I.

Radio broadcasting and telephony were organized in local or national regimes. In Germany, as in other European countries, these technologies were heavily regulated or monopolized by the state. Telegraphy and telephony on wires required a physical infrastructure that made regulation comparatively easy to enforce. Wireless telegraphy and wireless telephony, which in the 1920s became radio, did not require such a physical network of wires. Wireless waves did not stop at borders and could easily be intercepted without notice. In addition, wireless created a large international community of radio amateurs, who appropriated technology in their own ways and challenged and enlarged existing professional knowledge and concepts about the propagation of radio waves (Yeang 2013, 111–143). Wireless, and later radio, was nevertheless regulated both nationally and internationally, since it was seen as a matter of the state and the military. During and immediately after World War I, the military authorities of most nations made all ham radio activity illegal. Once this military ban had been lifted, state authorities assigned frequency bands and regulated equipment so that it would not interfere with other bands.

Technologies of sound recording, sound propagation, and sound amplification changed the relationship between sound, space, and time in a very physical sense. Methods of sound inscription, such as phonetic scripts or music notation, date back to antiquity, and mechanical sound reproduction by music boxes and player pianos also predates the nineteenth century. The mechanical sound recording of Édouard-Léon Scott de Martinville's phonautograph in the 1850s and the mechanical sound reproduction of Thomas Alva Edison's phonograph in 1877 were, however, regarded as something fundamentally new (Sterne 2003). Phonography was supposed to record and reproduce sound with the same degree of objectivity that was ascribed to photography for the recording and reproduction of visual images.[11] The phonograph conserved sound for eternity and allowed its reproduction wherever and whenever wanted, at least in principle.

Listening to these old cylinder recordings, we may judge their objectivity differently. We often have to guess the recorded words and might find it a mixed pleasure to listen to recorded music that is generally drowned by the noise of the scratching needle. Early photography might be evaluated similarly: not necessarily more true to the original than a drawing just because a machine produced it. This is not solely a modern criticism. The physicist and electrical engineer Heinrich Barkhausen, for example, argued in 1911 that human skills of listening were crucial to the reproduction of speech and music by the phonograph. According to Barkhausen, the reproduced sounds were severely distorted, and it was only the ear, in combination with human intellect, that made them intelligible.[12]

While phonography recorded for eternity, telephony and radio broadcasting transmitted sound almost instantaneously, much faster than the speed of sound itself, and over large distances. Telephony and radio created what John B. Thompson has called a "despatialized simultaneity" (J. B. Thompson 1995, 31–37). As Emily Thompson has shown, electroacoustics and other sound technologies changed the relationship between sound, space, and time on smaller scales as well. Around 1900, Wallace Sabine transformed the understanding of architectural acoustics. Before Sabine, sound quality in a concert hall or a theater was related to its geometrical shape and dimensions, but Sabine's reverberation measurements and formula shifted the focus from the geometry of the space to the sound absorption of construction materials (E. Thompson 2002). Sound-absorbing materials and electroacoustic amplification changed the acoustic qualities of spaces such as lecture halls, churches, motion picture theaters, concert halls, and sports arenas. Sound amplification systems and radio became, most infamously, important means for orchestrating the political mass rallies of the German National Socialist Party. The dispersed placing of loudspeakers made it possible to synchronize the masses effectively (Emde, Heinrich, and Vierling 1937).

From being something peculiarly local, sound had become universal and reproducible. Yet the sounds of language and music remained connected to individual and communal identities. Myles Jackson has written about the bourgeois scientific community in nineteenth-century Germany and the importance of singing folk songs for identity building in post-Napoleonic Germany (Jackson 2006). For the Nazis, the particular soundscape of political rallies helped to shape the communal experience of emotional and spiritual synchronization, which left no space for individual expression.

Old sounds and soundscapes did not necessarily die with the arrival of new ones. In the world of sounds, we can detect what David Edgerton has called the "shock of the old" (Edgerton 2007). All forms of classical, folk, and popular music are still performed today and enjoy unbroken popularity. Classical musical instruments are still built, by hand using old craft techniques as well as through mass production. When scientists and engineers introduced new electroacoustic musical instruments in the 1920s and 1930s, these were not very successful among musicians and composers (Donhauser 2007, 223–226). Sound location of artillery and aircraft played little role in the scientists' war effort during World War II, though the techniques were used and are in fact still used by the military today. To be sure, not all soundscapes survived. Silent movies accompanied by orchestras disappeared rapidly and almost completely within a few years of sound film's breakthrough in the late 1920s.

Scientific publications can have a long life span as well. Hermann Helmholtz's *On the Sensations of Tone* of 1863 is still in print and read by musicologists and musicians. In *Analog Days*, Trevor Pinch and Frank Trocco retell a fascinating story by Malcolm Cecil of the synthesizer band Tonto, set in the early 1970s. Cecil owned a copy of Helmholtz's *On the Sensations of Tone*, which he had studied in detail. When he wanted to produce bell sounds on his synthesizer, he remembered that Helmholtz had analyzed the bell of Kiev and written down its harmonics:

So we dialed up the harmonics from the great bell of Kiev, exactly as Helmholtz had written ..., fed them into the mixer, put them through the filter, put the envelope on there that we'd already figured out, pressed the key, and out came this bell. I'm telling you, it happened. It was unbelievable! We were hugging each other, dancing around the studio. "We did it, we did it, we did it, we did it!" (Pinch and Trocco 2001, 181)

As with all anecdotes, this has to be read with some caution—it was not the great bell of Kiev but the great bell of Erfurt Cathedral for which Helmholtz noted the harmonics, and Helmholtz did not do the analysis himself but cited the organist Carl Anton Gleitz (Helmholtz [1863] 1875, 118). The story nevertheless documents the continuing relevance of Helmholtz's *On the Sensations of Tone*, not only for views of classical music but for the production and performance of modern music, far beyond what Helmholtz might have imagined.

1.3   Sound and Power: Acoustics as an Imperial Science

Let us turn now to a more traditional understanding of geography and its intersection with acoustics. In view of the relationship between sound and various forms of power, acoustics can be regarded as an imperial science. Producing and controlling sounds has had implications for exercising political, military, and economic power. In 1901, the geographer Siegmund Günther presented a two-part paper entitled "Akustisch-geographische Probleme" (Acoustical-geographical problems) to the Bavarian Academy of Sciences and Humanities (Günther 1902). In it, he discussed sound phenomena originating in physical geography, such as sounding or singing sands, singing valleys, singing forests, singing rocks, and sounds emerging from streams and waterfalls. He drew on reports by Alexander von Humboldt, Charles Darwin, and other less well-known travelers. Günther reported most enthusiastically about the harmonious consonant tones that natural phenomena produced. For Günther, nature was literally singing. It was not songbirds and other singing animals, but singing landscapes that fascinated him. Singing or not, nature sounded different across the earth, and soundscapes were peculiar to landscapes and other geographical characteristics.

Günther mentioned the influences that human activity could have on soundscapes, such as the effects of deforestation on singing forests and of other manmade changes to the landscape. In explorers' travelogues, they generally reported less on the soundscapes produced by the landscapes, as discussed by Günther, and more on the cultural sonic expressions of local populations. Traveling and colonial scientists studied first the speech and music of the local populations in the shape of ethnomusicology, phonetics, and linguistics (Ames 2003; B. Lange 2011; Stangl 2000; Stumpf 1908). Here, acoustic research had several overlaps with the colonial projects of the European imperial powers. For example, soundscapes of nature and culture changed with increasing colonization, industrialization, and urbanization. The industrial landscapes and soundscapes of modern metropolitan cities such as Berlin, London, New York, Paris, or Buenos Aires had little in common with Günther's singing forests and valleys.

With the global spread of sounds and sound technologies around the turn of the twentieth century, ethnographers and others expressed the worry that local languages and local traditions of music would be lost. Colonization and modernization caused a disruption of indigenous tribal structures and culture, if not the extinction of whole tribes, which many believed would lead to an increasing uniformity of language and music. Carl Stumpf, the founder of the Berlin Phonogram Archive, argued that it was the duty of the new German Empire to exploit the colonies not only economically but also scientifically (Stumpf 1908, 245)—the culture of the native population had to be studied in all aspects and with scientific accuracy. Phonograph recordings, wrote Stumpf, would facilitate such scientific accuracy for ethnographic studies of speech and music. The phonograph became the machine that could preserve for eternity indigenous language and music before the speakers and their culture vanished. What Stumpf forgot to mention was that the colonizers exterminated indigenous populations and their cultures deliberately, as in the German Empire's genocide during the Herero Wars in German South West Africa between 1904 and 1908 (Zimmerer and Zeller 2004).

Language and music were connected to ideas of cultural supremacy and domination, nationalism, and the claim to a certain type of modernity, for both language and the performance and consumption of music are inextricable from individual and collective identities. What was at stake for the ethnographers and musicologists when they took the phonograph to foreign parts of the world was the idea of the universality and supremacy of classical European music as high culture (Hui 2013, 125–144). After the introduction of the phonograph in 1877, anthropologists, linguists, and musicologists traveled around the world to record language and music from different cultures, analyzing the various musical systems and comparing them with systems of European classical music. The comparative study of musical systems showed the musicologists that concepts and systems of European classical music could not be easily transferred to the music of other cultures, and that notions such as harmony and disharmony, consonance and dissonance could not be established as scientific facts in the physics and physiology of human hearing but were subject to psychological and cultural factors. Indeed, the ideals of classical European music were challenged not only in the colonies but also at home in Europe, by the Italian futurists. Luigi Russolo's manifesto *L'arte dei Rumori* (*The Art of Noises*) of 1913 broke down the boundary between musical tones and noise (Russolo [1913] 1986). It also effectively forecasted the symphony of the industrial war.

One might imagine that the exigencies of World War I would have put a stop to the research of ethnographers and linguists. To the contrary, as Britta Lange has shown, German prisoner-of-war camps during World War I enabled these researchers to extend their phonetic studies to captive soldiers drawn from the colonies. The camps became huge laboratories where scientists recorded speech and songs of soldiers from around the colonized world. The prisoners of war and other subjects of these studies, however, were not given a voice to speak for themselves. They were measured and characterized with scientific precision. Colonial hierarchy reflected scientific hierarchy. It was European scientists who studied the colonized. Even when the captives from the colonies of the imperial armies spoke to be recorded on the phonograph for linguistic studies, they remained subaltern, as nobody listened to their stories (B. Lange 2011).

Not only the colonized, but also the colonizers were imprisoned on foreign soil during World War I. The German operators of the Kamina wireless station in the German colony of Togoland (Togo) were taken prisoner after British and French troops invaded the colony. Telegraphy, both on wires and wireless, was crucial to imperial and colonial agendas, where the political and geostrategic interests of the empires merged with the commercial interest of the colonizers. With the telegraph and wireless, European powers set up a global communication network. The British telegraph network has been described as the "nervous system" of the British Empire (B. Hunt 1991, 54), and Aitor Anduaga has shown the entanglement of wireless research with geostrategic, geopolitical, and economic interests in the British Empire (Anduaga 2009). In a similar manner, German scientists, statesmen, and military leaders acted on the crucial importance of telegraphy and wireless for the young but ambitious German Empire (Friedewald 2002). Gesellschaft für drahtlose Telegraphie System Telefunken (Telefunken), created as a German rival to the Marconi Wireless Telegraph Company, had to overcome the shortcomings of telegraphy on wires. Because telegraphy on wires could not be used to communicate with ships at sea, establishing a worldwide network of wireless telegraphy was essential for communicating with the expanding military and merchant navy.

But there was more to wireless. Telefunken's wireless agenda was truly imperial: cables had to cross other countries and colonies to reach the scattered German overseas possessions. The German colonial empire was spread out, with no land connection to the Reich, and telegraph connections could easily be cut by an enemy in the case of war, as happened in the early days of World War I. Yet, the wireless station of Kamina did not survive long. Having only gone into operation in July 1914, it was destroyed by the German operators themselves on 24 and 25 August 1914 prior to their surrender (Esau 1919). The operators were imprisoned in Gaya, in the French colony of Niger. Needless to say, the German prisoners of war were not subjected to phonetic studies. Back in Germany, the anything-but-voiceless German colonizers spoke out in an article in the *Telefunken-Zeitung*. Not hiding his racism, engineer Carl Doetsch complained bitterly that the French had not treated German prisoners of war as civilized Europeans. The French Army had imprisoned them "in the interior of Africa at such an unhealthy place, guarded by negroes and living like negroes" (Doetsch 1920, 36).

The European colonizers, scientists or otherwise, insisted on their superiority and exceptional status. But, as Marwa Elshakry has pointed out, native scholars in Egypt, China, and India did not necessarily see scientific knowledge and technologies as specifically "Western" or foreign to their own scientific or technological understanding (Elshakry 2010). Claims to the universality of knowledge are not unique to "modern science" as it developed in Europe, but are equally prevalent in other cultures and systems of knowledge. New knowledge about sound, new sound technologies, and new sounds originating from Europe were not necessarily seen as exceptional or alien to existing systems of knowledge, technologies, or soundscapes in other parts of the world. Acoustics research in the tradition of Helmholtz was an academic activity that an Indian physicist like Chandrasekhara Venkata Raman could identify with and use as a source of inspiration for a scientific identity and career:

It was my good fortune, while a student at college to have possessed a copy of his [Helmholtz's] great work *The Sensations of Tone*. … I discovered the book myself and read it with the keenest interest and attention. It can be said without exaggeration that it profoundly influenced my intellectual outlook. For the first time I understood from its perusal what scientific research really meant, and how it could be undertaken. (Raman n.d., cited in Ramaseshan 1988, x)

Raman, who received the Nobel Prize for Physics in 1930 for his work on what is now known as Raman scattering, was only one of several Indian physicists working on acoustics at the time. As the historian Gyan Prakash has shown, Indian scholars and other members of the elite could easily connect to the project of universal science (Prakash 1999, 8–9). Raman, as a Tamil Brahmin, belonged to the Indian scholarly elite and was trained in Indian classical music. Raman and other Brahmins related to and practiced Indian classical music in ways not unlike the German bourgeoisie's relationship to and practice of German classical music. It is therefore no surprise that Raman was attracted by Helmholtz's program of founding the aesthetics of music on the science of sound and its sensation. Raman experienced the political, social, and economic realities and limitations of practicing science in the colonial setting of British India; nevertheless, his access to education and scientific institutions, as well as sharing a common set of values with European scientists, enabled him to participate in scientific enterprise. Raman contributed the chapter on musical instruments to the volume on acoustics in the monumental *Handbuch der Physik*, published in 1927 (Raman 1927). A professor of physics at the University of Calcutta, he was the only author of the volume who did not work in Germany and one of the few not based in Berlin.

The transnational economic and cultural importance of acoustic technologies and knowledge became particularly evident with the breakthrough of sound film in the late 1920s. Sound motion pictures came from the United States to Europe with the films produced by Warner Bros and the Vitaphone sound system, developed and marketed by Westinghouse, a subsidiary of American Telephone and Telegraph (AT&T). The German electrical industry, represented by Allgemeine Electricitäts-Gesellschaft (AEG) and Siemens, and the film production companies Universum Film AG (UFA) and Tonbild Syndikat AG (Tobis) acted fast to develop their own sound motion picture system so that they would not be swept away and lose their national and international markets.

The economic significance of sound motion pictures was overshadowed by its cultural significance, we might argue. With sound film, as well as radio and the recording industry, acoustics became implicated in cultural transformation. From a science of music as high culture, acoustics became a science of the cinema as mass culture. Sound motion pictures embodied Walter Benjamin's 1936 *Das Kunstwerk im Zeitalter seiner technischen Reproduzierbarkeit* (Work of art in the age of mechanical reproduction). Combining photography and phonography, sound motion pictures made the quest for original and reproduction meaningless for both (Benjamin [1936] 1963). The music educator Leo Kestenberg published the collection *Kunst und Technik* (Art and technology) in 1930, shortly after the breakthrough of sound motion pictures in Germany. This work brought together a variety of authors, including philosopher Ernst Cassirer, sociologist Paul Honigsheim, physicist Erwin Meyer (head of the Acoustics Department of the recently founded Heinrich Hertz Institute for Oscillation Research), composer Ernst Krenek, and Heinrich J. Küchenmeister, one of the entrepreneurs of sound film in Germany.[13]

In *Kunst und Technik* and elsewhere, cultural values and standards around modernity and other issues were transported together with electroacoustic technologies and soundscapes. Honigsheim saw jazz, film music, and radio as the new forms of musical expression. Jazz, he wrote, satisfied the ear because it corresponded to the noises of the street and the factory, while film music was meant to drown the noise of the projection apparatus (Honigsheim 1930, 74). Honigsheim's comments take us directly to the noise abatement debates and the anti-noise leagues that formed in Berlin and in other metropolitan areas, such as New York and London (Bijsterveld 2008).

1.4   Concepts and Significance of Noise in Acoustics

Noise is a central concept in acoustics. It concerns distinguishing certain kinds of sounds from others. Historians have written about noise in the interwar period in the context of the anti-noise movement, public hygiene, and the modern city (Braun 1998; E. Thompson 2002, 115–168; Bijsterveld 2008). I offer a different view, arguing that at least three historical developments altered the concept and significance of noise within science: the deployment of acoustics and acousticians in World War I, the rise of electroacoustics and media technologies such as telephony and radio broadcasting, and the emergence of comparative musicology. Notions of noise associated with the measurement process and information theory have spread beyond sound to virtually all fields of science and engineering, even the social sciences. Not surprisingly, we now find not one but several different and even contradictory concepts of noise in science and engineering. How did this happen?

Scientific concepts of noise have changed over time, but they have also varied in relation to the particular research field or technical application in which they are embedded. As a consequence, we have to historicize concepts and notions of noise in acoustics and understand the different contexts in which they originated. In this book, I go beyond the noise abatement debate and discuss how scientists defined and framed noise from the 1860s to the 1930s. In doing so, I want to challenge and expand some of the histories of noise that have been prevalent so far. If we compare the German scientific discourse about noise with the discourse in the English-speaking world, we discover that at least three different notions have emerged in the German language, *Lärm*, *Geräusch*, and *Rauschen*, which all translate into "noise" in English. As I will show, these distinctions in the German language had consequences, at least in the German debate about noise in the interwar period. To briefly introduce the three concepts:

1.   Geräusch (noise as nonperiodic sound): In Hermann Helmholtz's influential *Die Lehre von den Tonempfindungen als physiologische Grundlage für die Theorie der Musik* (*On the Sensations of Tone as a Physiological Basis for the Theory of Music*) of 1863, he divided sounds into musical tones and noises, claiming that all musical tones were harmonic (or periodic), while noises were disharmonic (or nonperiodic). This definition was adopted in Lord Rayleigh's *The Theory of Sound* of 1877.

2. Rauschen (electrical noise): Electrical noise emerged as an analog to acoustic noise with the advance of electroacoustics in the 1920s. Electrical noise was generated in electric circuits, for example, in microphones and radio tubes. These sounds were often also described as buzzing or humming (*Summen* in German), as they could be both harmonic and nonharmonic.

3. Lärm, *Störschalle* (noise as nuisance): In this definition, every unwanted sound was noise, whether harmonic or disharmonic. While long recognized as a problem in the public sphere, scientists first started to deal with noise as nuisance during World War I and the noise abatement initiatives of the interwar period.

Noise could be defined as the opposite of silence or of a certain sound signal. The various discourses about noise participated in the transformations of the public and scientific soundscape from the late nineteenth to the midtwentieth century. Three main forces drove the changing concepts of noise in this period:

1. Changes in understanding music as a result of comparative musicology and mass culture, moving away from classical European music as high culture and its theoretical framework

2. The industrialization of urban life and World War I as an industrial war: World War I turned acoustics away from music as its main frame of reference and toward the soundscapes of the battlefield

3. The emergence and growth of electroacoustic media technology, especially telephony, radio broadcasting, and sound motion pictures; with electroacoustics and electroanalog thinking, noise, which originated as a classification of sounds in acoustics, became a category of electric systems and electrical engineering

1.5 Electroacoustics and Analog Thinking

The transformation of acoustics into electroacoustics was central to the reshaping of the science of sound in the first half of the twentieth century. In addition to science, electric recording, transmission, manipulation, and amplification of sound have defined much of our common aural experience ever since. With the rise of electroacoustics, the way music was produced, consumed, and understood was transformed. The history of electroacoustics has thus far been approached predominantly as a history of technologies such as telephones, microphones, loudspeakers, and electric amplification. These were coproduced with new technologies of mass media, particularly radio broadcasting and sound film, at the same time as they entered the research laboratory (F. V. Hunt 1954; Beyer 1999, 177–186; E. Thompson 1997, 2002, 229–293). My focus is a different one: I argue that this transformation of acoustics into electroacoustics went far beyond electric technology and led to a conceptual redefinition of sound. To understand these transitions, we must consider both the development of electric technologies and the changing understanding of electrodynamics—especially electric oscillations and electric circuit design.

The electroanalog field became a new language, a new way of thinking and talking about sound in the twentieth century that used electricity as its core metaphor. Its practices were shaped by the circuit diagram, as well as through linguistic and mathematical expressions. Circuit diagrams became firmly established in the 1930s, part of a whole array of abstract visual concepts, such as flowcharts and thermodynamic cycles, developed by scientists and engineers. Within the framework of these visual concepts, technical problems could be formulated, analyzed, and solved.[14] Just like Christopher Polhem's mechanical alphabet and Franz Reuleaux's machine grammar, developed in the eighteenth and nineteenth centuries, circuit diagrams conveyed their own semiotics and grammar. These consisted of standardized symbols for standardized electric components and a network of idealized electric connections (Ferguson 1992, 137–147). While Polhem's mechanical alphabet and Reuleaux's machine grammar are almost forgotten, circuit diagrams have been very successful and still constitute an important and widespread scientific and engineering practice. Though engineers and scientists could easily envision an actual electric circuit from the diagram and produce it in the workshop, the relationship between the drawing and the material circuit was not straightforward. Circuit diagrams were not about the materiality of the circuit, but about its operation. In contrast to mechanical drawings, circuit diagrams were meant to show functional relations, not spatial arrangements.[15]

This abstract and conceptual nature of circuit diagrams was especially prominent in electroacoustics. Circuit drawings could represent actual electric circuits such as amplifier circuits, microphone and loudspeaker wiring, or electric filters. They could also represent acoustic systems that were partly or entirely nonelectric. This is well illustrated by an example from research in sound insulation in buildings at the Norwegian Institute of Technology in the 1930s. In figure 1.1, an equivalent circuit diagram illustrates the transmission of sound through a double wall. While the upper drawing shows the wall construction with the incoming and transmitted sound, the lower equivalent circuit diagram represents the wall as an oscillating system with dampening.[16]

What is truly remarkable in this representation of the double wall by an equivalent circuit diagram is that there is nothing electrical about the wall or the sound traveling through it. By the 1930s, electric circuit diagrams, even when representing nonelectric systems, had become a lingua franca of the new acoustics. Moving from the electric to the acoustic system was an act of translation. Electric oscillations with the same waveforms as acoustic vibrations could be described using the same mathematical equations, by substituting the equivalent electric variables for the acoustic variables. Acoustic variables such as force, speed, displacement, mass, and elasticity were replaced by such electric variables as tension, current, charge,

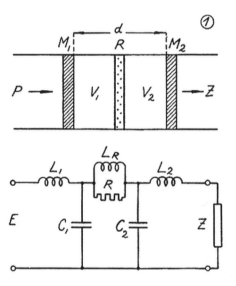

Figure 1.1

The electric equivalent circuit diagram of a double wall as an oscillating system with dampening (Berg and Holtsmark 1935b, 75).

self-induction, and capacity. The electric systems then became mathematically identical representations of the acoustic systems.[17]

The use of analogies between different physical phenomena was not uncommon throughout the nineteenth century. Analogies served didactic purposes and linked apparently separate phenomena in order to achieve the ultimate goal of a unified physical science. Analogies between electric and acoustic features were frequently used, but so were analogies with light, heat, and other phenomena. The use of the term "electroacoustics" itself can be traced back to 1903, when it first appeared in the German literature (Hartmann-Kempf 1903). The timing coincided with the rise of an electromagnetic worldview that sought to replace mechanical conceptions with electromagnetic ones as the foundation of the physical sciences. The electromagnetic regime eventually failed as a worldview, but it brought the development of electromagnetic representations and electric technology to the forefront of the physical sciences and engineering.

Frederick V. Hunt began his extensive historical introduction to electroacoustics with the proclamation that the field was as old as thunder and lightning, which is, for the scientist, a natural electroacoustic phenomenon (F. V. Hunt 1954, 1). For our purposes, however, it will be sufficient to trace the history of the entanglement of acoustics with electric technology and electrodynamic theory back to the middle of the nineteenth century and its two most influential works on acoustics, those by Hermann Helmholtz and Lord Rayleigh.

1.6   Transformations of Science and Sound from the Fin de Siècle to the Interwar Years

The chronological narrative of the history of acoustics covered in this book is divided into four periods, each presented in a separate chapter. The first period stretches from the 1860s to the outbreak of World War I. The second chapter covers World War I. I have then divided the interwar period into the Weimar Republic and the beginning of the Third Reich, from the Nazi seizure of power in 1933 to the outbreak of World War II in 1939. This division highlights that the key actors and institutions of my narrative of the interwar period were located in Germany, and that the Nazi seizure of power profoundly affected the acoustics research community. The concluding chapter comes back to some of the epistemological questions related to the transformation of acoustics during the four periods.

Chapter 2 opens the narrative by presenting and contextualizing acoustics research as set out by Hermann Helmholtz in Bismarck's Germany and Lord Rayleigh in Victorian Britain in the 1880s and 1890s. Whereas Helmholtz wrote as a physiologist and a physicist for aestheticians and musicologists, Rayleigh wrote for mathematically trained physicists. Their frame for acoustics as a science of musical sounds, I argue, was nevertheless the same. While Helmholtz and Rayleigh developed visual and other mechanical instruments for sound measurement, the musically trained ear remained the ultimate reference for sound observations. The questions Helmholtz posed of consonance and dissonance and combination tones continued to define acoustics research until World War I, but the search for answers moved from physics and physiology into experimental psychology and comparative musicology. The instrument maker Rudolph Koenig and the philosopher and psychologist Carl Stumpf were Helmholtz's harshest critics in this debate. Whereas Helmholtz located the foundations of aesthetic sound perception in physics and physiology, comparative musicologists now drew on Charles Darwin's theory of evolution as the foundational scientific principle in the development of music.

At the same time, physicists moved away from acoustics and toward electrical research, especially after Heinrich Hertz's groundbreaking experiments on the propagation of electric waves. With the focus on electromagnetism, electrical engineering, and electric technology such as the telephone and the microphone, a new kind of acoustics emerged: electroacoustics. What characterized electroacoustics was the strong connection between acoustics, wireless (later radio) technology, and electroanalog thinking. Hermann Theodor Simon, the director of the Institute for Applied Electricity at the University of Göttingen, was one of the scientists who brought these different fields together. Simon developed a research program around the electric arc as an experimental system. Among Simon's students working on different aspects of the electric arc were Heinrich Barkhausen, Hugo Lichte, and Karl Willy Wagner, who all played important roles in the acoustics research community of the interwar period. With the electric arc, Simon and his students placed acoustics within a fundamentally electrical frame, and within the electromagnetic worldview.

Chapter 3 follows the transformation of acoustics during World War I. The first large industrial war led to an industrialization of acoustics. Acousticians turned away from music as high culture and to the complex sounds of the battlefield. Acoustic knowledge and technologies became important for the sound location of artillery and aircraft and the detection of U-boats. For the first time, acousticians became interested in industrial noises and noise abatement. Notions of signal on one hand and noise on the other were determined not by physical abstractions but by the necessities of warfare. Until World War I, almost all sound measurement was done by the human ear, but the human as an observer became problematic when it came to precise sound and time measurement. Physicists worked on automating measurement, using what they called objective systems, in which the ear was replaced by a microphone and a recording apparatus. Lucien Bull, Lawrence Bragg, and William S. Tucker achieved the automation of artillery ranging for the Franco-British sound-ranging system. Not only did the war advance automated sound measurement, but Tucker's hot-wire microphone also acted as a filter, registering certain sounds while ignoring others.

Equally important for the electroacoustics research agenda was the development of radio. Wireless telegraphy opened up new possibilities for military communication. Radio tube transmitters enhanced frequency selectivity dramatically and enabled less bulky equipment for aircraft and mobile units. Wireless telephony took over from telegraphy. Scientists in the military and industry put great effort into the improvement of radio tubes, which became the key element of circuit design and of electroacoustic technology. The strong link between radio, circuit design, and electroacoustics led to the transfer of the concept of noise, originally a concept from acoustics, to radio and electrical engineering.

Chapter 4 concerns the development of acoustics in the Weimar Republic. The Versailles Peace Treaty led to a rapid demilitarization of acoustics research and development. Yet, the employment of physicists in technoscientific projects during the war had firmly established a close connection between science and industrial development. *Technische Akustik* (technical acoustics) became an archetype of technical physics, with its importance for the development of radio broadcasting, sound motion pictures, sound recording, and sound amplification systems in the 1920s and 1930s.

The new generation of technical physicists were to be found most commonly not at universities but at the German institutes of technology, the *Technische Hochschulen*, in industrial research laboratories and in research and development institutions located at the intersections of science, state, and industry. At the Technische Hochschulen, the advance of physicists into industry led to tensions with engineers and the need for lines of demarcation between physics and engineering. Heinrich Barkhausen became a pioneer of low-current electrical engineering as a professor at the Technische Hochschule Dresden. In the interwar period, he followed a research program in electroacoustics within electrical engineering, based on the concept of *Schwingungsforschung* (vibration and oscillation research). At the Technische Hochschule Munich, the physicist Jonathan Zenneck also connected his acoustics research more or less seamlessly with his main research field of high-frequency radio waves.

The Siemens and Halske company and the Allgemeine Electricitäts-Gesellschaft (AEG) were the two main players in acoustics research in the German industrial landscape. Hans Riegger developed loudspeakers and large amplification systems for Siemens, drawing on his experience from underwater acoustics during the war. Siemens tested Riegger's loudspeaker systems, as well as those of Erwin Gerlach and Walter Schottky, at the opening of the Deutsches Museum in Munich at its new premises in 1925. AEG launched its research institute in 1928, and Hugo Lichte led the development of the German sound-on-film system as head of its Acoustics Department.

A new type of research institute emerged in the Weimar Republic, which created a link among the Technische Hochschulen, industry, and state institutions. In 1930 the Heinrich-Hertz-Institut für Schwingungsforschung (Heinrich Hertz Institute for Oscillation Research) opened its doors in Berlin. Its founder and director, Karl Willy Wagner, had first set out plans for a research institute with representatives from the German national broadcasting corporation, the electrical industry, and the Technische Hochschule Berlin-Charlottenburg in 1926. Wagner adopted Barkhausen's concept of Schwingungsforschung as the overarching research program for the new institute. Erwin Meyer, a student of Erich Waetzmann's in Breslau, was appointed to head the Acoustics Department of the Heinrich Hertz Institute and soon became one of Germany's most recognized acousticians.

The Institut für Schall- und Wärmeforschung (Institute for Sound and Heat Research) in Stuttgart had a similar structure but followed a very different research program and trajectory. The work of this institute was directed not at electroacoustics and new media but at the testing of new construction materials and methods. Wagner's agenda for the Heinrich Hertz Institute was to compete internationally. The Institute for Sound and Heat Research, in contrast, provided acoustic expertise and practices for local consumption.

In chapter 5, I examine the National Socialist rise to power in 1933 and its profound impact on the acoustics research community. Technical physics in general, and especially technical acoustics, squared well with the Nazi agenda, which dismissed the idea of pure science as an end in itself and adopted the concept of a science that served technological goals. Technical acoustics offered everything that National Socialists wanted "German Physics" to be: experimental, applied, and relevant to the military. Acoustic knowledge and technology was decisive for the orchestration of Nazi political rallies and propaganda shows. The name of Heinrich Hertz, who was of Jewish ancestry, was removed from the Institute for Oscillation Research. Wagner and other scientists were dismissed in political and anti-Semitic purges and in personal retaliation. The advent of the Nazi regime also enabled Hermann Reiher, the director of the Institute for Sound and Heat Research, to be appointed as professor of technical physics in Stuttgart. In line with the military mobilization of German science and society at large, the acoustics research agenda was quickly remilitarized, building on experiences and practices from World War I. Johannes Stark, one of the leading representatives of "German Physics," was appointed director of the Physikalisch-Technische Reichsanstalt in 1933. Stark quickly established a group on acoustics and appointed Martin Grützmacher, who would work on acoustic torpedoes during World War II. Erwin Meyer of the Institute for Oscillation Research headed efforts to make German submarines invisible to sonar. With research in pure physics under attack in Nazi Germany, Carl Ramsauer became the advocate for reestablishing the hierarchy between pure and applied physics, as well as for the necessity of a research agenda in pure physics.

In chapter 6 I look back on the new acoustics that emerged in the first decades of the twentieth century. What was the epistemological basis of the new acoustics? What were its scientific practices? Who were the new acousticians, where did they come from, and what was their identity? Finally, how does the history of technical acoustics and electroacoustics fit into the existing historiography of physics in the interwar period and the apparent dichotomy between science and engineering?

The acoustic laboratory and what it meant to measure sound was transformed beyond recognition in the interwar period. The traditional acoustic instruments used by Helmholtz, Koenig, and others disappeared from research; electroacoustic measurement technology based on radio took over. Experimenters needed electroanalog thinking and practices from electrical engineering rather than a musically trained ear. The range of sounds that were measured and classified had widened, and the new concepts of signal and noise had entered the laboratory. With electrical engineering and communication technology, those very concepts of signal and noise also spread to other fields of science.

As part of technical physics in the interwar period, technical acoustics was located in the boundary region between physics and engineering. Even though the boundary was permeable and had to be constantly renegotiated, physicists were always eager to explain that technical physics was not engineering. In Suman Seth's study of the Sommerfeld school, he distinguishes between "physics of problems" and "physics of principles" as approaches to physics in the first decades of the twentieth century (Seth 2010). In Seth's terms, technical acoustics represented a physics of problems rather than a physics of principle; its problems arose specifically from technology and industry. The separation into physics of problems and physics of principle, however, misses one important point. The new acoustics research in the interwar period was guided by rather strong overarching principles. Electroanalog thinking and concepts of signal and noise belonged to the conceptual toolbox of the acoustician. Closely related to analog thinking was the larger principle of *Schwingungstechnik*. The understanding of oscillations and vibrations was fundamental to all fields of physics, including the new fields of quantum physics and relativity. At the same time, acousticians from Hermann Helmholtz in the 1860s to Ferdinand Trendelenburg in the 1930s agreed that the science of acoustics should not be reduced to the physics of elastic bodies, but was inseparable from human hearing, and from speech and music as cultural expressions.

2 The Electrification of Sound: From High Culture to Electropolis

2.1 Acoustics in the Age of Empire

The works of Hermann Helmholtz and John William Strutt, 3rd Baron Rayleigh, were the benchmark of acoustics research in the nineteenth century. Helmholtz and Rayleigh were masters of both mathematical analysis and experimental investigation. Helmholtz's *Die Lehre von den Tonempfindungen als physiologische Grundlage für die Theorie der Musik* of 1863 (translated in 1875 as *On the Sensations of Tone as a Physiological Basis for the Theory of Music*) and Rayleigh's *Theory of Sound* (2 vols., 1877 and 1878) were widely read throughout the nineteenth and twentieth century and are still in print today.

Until the outbreak of World War I, acoustics remained the quintessence of nineteenth-century bourgeois science, a world in which the sensation of sound and the experience of music were interchangeable, and bourgeois music, bourgeois science, and bourgeois society were coproduced. The masterpieces of nineteenth-century acoustics—both scientific knowledge and instruments—were based on a fundamentally mechanical understanding of sound. This mechanical understanding and standardization of sound stood in a relationship of tension with music's free spirit and cultural values, and with the idea of the virtuoso performer (Jackson 2006; Hui 2013, xiii).

After Helmholtz and Rayleigh, discourses and developments in acoustics took two main directions. The emerging fields of experimental psychology and ethnomusicology challenged Helmholtz's understanding of the relationship between the science of sound and the aesthetics of music. Ethnomusicology emerged as part of the colonial ambitions of the German Empire, employing a Darwinian rather than a physical and physiological notion of music. The second direction was an electrification of acoustics, in terms of electroacoustic technologies and the development of electromagnetic theory as the new paradigm for physical research. The concept of electroacoustics appeared around the turn of the century but was fully developed only during World War I. Taking Helmholtz and Rayleigh as a point of departure, in this chapter I trace the transformations of acoustics under the influence of experimental psychology, comparative musicology, and the increasing use of electric technologies and analogies in acoustics research. I conclude by placing acoustics within the electrical industry and electromagnetic worldview that dominated the end of the long nineteenth century.

Hermann Helmholtz was about forty-two years old when he published *On the Sensations of Tone*. He carried out his work on acoustics while a professor of anatomy and physiology in Bonn (1851–1858) and Heidelberg (1858–1871). This was also when Helmholtz completed his monumental *Handbuch der physiologischen Optik*, which appeared in three volumes in 1856, 1860, and 1866 (Helmholtz 1867, v). The Heidelberg period, with the work on physiological optics and acoustics, marked the end of a phase in Helmholtz's scientific career. When Helmholtz moved to Berlin in 1871, he took on the prestigious chair in physics, succeeding Gustav Magnus, at Friedrich Wilhelm University. Helmholtz now changed his research agenda from the physiology of sensations to electrodynamics. He had already carried out research on electricity in his work on nervous excitation in the 1850s. In the later period, Helmholtz chose a more comprehensive approach, seeking to solve fundamental questions and achieve a synthesis of electrodynamic theory. Before his move to Berlin, he had aimed to base the principles of physiology upon the laws of physics; once in Berlin, Helmholtz set his sights on the foundations of physics itself (Kaiser 1993).

The year Helmholtz arrived in Berlin, 1871, also marked the beginning of a new era in German history. Berlin became the center of power in the newly unified German Empire. After the Wars of German Unification between 1864 and 1871, Prussia emerged as Germany's new powerful leader, and in the Franco-Prussian War of 1870–1871, Prussia and its allies defeated the Second French Empire. In the aftermath, the German Empire ascended to the status of a major industrial and economic power in Europe. Industrialization, economic growth, and educational reforms led to an expansion of the physical institutes of the German universities and Technische Hochschulen (university-level technical colleges).[1] When Helmholtz moved into the newly built Physical Institute of the Friedrich Wilhelm University in 1876, it was the flagship of these new institutions.

Helmholtz's focus on electrodynamics and his rising influence within the scientific community was in tune with electrification and the growing electrical industry in Berlin. The aspiring metropolis became known as Electropolis, the prototype of a city defined by electricity (König 1995, 32). There was now a "national meaning" to all scientific work, and "science and politics were to go hand in hand in the new Reich" (Cahan 1989, 11). When Wilhelm II came to power in 1888, the German Reich heightened its colonial and imperial ambitions. Science had an important role to play in the expansion and government of the new empire, which founded the Hamburg Colonial Institute (Hamburgisches Kolonialinstitut) in 1908 to train administrators for the colonies. Wilhelm II pushed to build up a strong German Navy that could rival the British Royal Navy in controlling the oceans. Nonetheless, it was primarily the rise of the German industry that established a direct link between science and the Reich, and only secondarily the military and colonial project. Whereas networks of telegraphy had given rise to a modest electrical industry in the mid-nineteenth century, electrification, electric lighting, and heavy current machinery drove the growth of the industry in its later decades. The relationship between physics and electrical engineering was not as straightforward as that between chemistry and chemical engineering. Chemistry had a longer tradition of close connections between research and industrial production, especially in the synthetic dye industry. Electrical engineering in Germany, in contrast, grew out of mechanical engineering and industrial practice as much as it grew out of physics (König 1995). The relevance of scientific methods, as opposed to practical experience alone, for electrical engineering was contested, not only in Germany but also in Britain and the United States.[2]

In this climate, the industrialist Werner von Siemens, one of the founders of the electrical company Siemens and Halske, led the initiative to establish the Physikalisch-Technische Reichsanstalt (Imperial Institute of Physics and Technology), a national institute for metrology and standards. The mission of the Reichsanstalt was twofold: to conduct fundamental research in metrology, and to carry out calibrations and testing for German industry and public institutions. When Helmholtz, who was Siemens's close acquaintance for many years, was appointed first president of the Reichsanstalt in 1887, he became a symbol of the proximity of science, technological progress, and industrial might.[3] The model of the Reichsanstalt was copied in other countries, prominent examples being the National Bureau of Standards in Washington, DC, and the National Physical Laboratory in Teddington (Cahan 1989, 3). According to Leo Königsberger's monumental biography, Helmholtz rose to the status of a "Bismarck of the sciences" (Königsberger 1902–1903, 3:97)—he was arguably the most powerful person in German physics, and he outlived Otto von Bismarck by four years.

None of this could have been predicted in 1863. Helmholtz's research in acoustics was located in the German tradition of Ernst Florens Friedrich Chladni, Wilhelm Weber, Georg Simon Ohm, and August Seebeck (Jackson 2006), but if the publication of *On the Sensations of Tone* ended a chapter in Helmholtz's scientific biography, it broke new ground in acoustics research and dominated the debates for the next fifty years, until the outbreak of World War I. Helmholtz addressed *On the Sensations of Tone* specifically to musicologists and aesthetic theorists. Regardless of this, he was also widely read by physicists, physiologists, and technology entrepreneurs.

Together with Helmholtz, William Strutt (later Lord Rayleigh) had a profound and lasting impact on acoustics and its development. If there was one scientist in Victorian Britain who had a standing similar to Helmholtz's in Germany, it would be William Thomson, the later Lord Kelvin—but Rayleigh did not lag far behind Kelvin as one of the most influential physicists of the Victorian era. He was about twenty years younger than Helmholtz and Thomson, and a rather different character, embodying the British gentleman scientist. While Helmholtz and Thomson were elevated to nobility, Strutt inherited his title after the death of his father in 1873.[4] As a mathematical physicist, he was both a product and a champion of the Cambridge system of the Mathematical Tripos. Rayleigh did not limit himself to mathematical theory but made significant experimental contributions in acoustics and in other fields. Rayleigh started to work on *The Theory of Sound* during a houseboat trip on the River Nile in 1872, which he undertook for health reasons. *The Theory of Sound*, a mathematical tour de force that applied differential calculus of vibration theory to acoustic vibrations, was probably Rayleigh's most influential contribution to science. Physicists could apply its mathematical methods to all kinds of physical problems beyond the field of acoustics. Specifically, Rayleigh's mathematical treatment of acoustic wave propagation became useful for understanding James Clerk Maxwell's rather obscure *Treatise on Electricity and Magnetism* of 1873 (Warwick 2003). Rayleigh agreed to succeed Maxwell as Cavendish Professor of Physics after Maxwell's death in 1879. In 1884, he left Cambridge and returned to his estate at Terling, where he conducted scientific research in his private laboratory.

Emphasizing mechanical understandings of acoustics, the technologies of acoustic measurement at this time were based on precision mechanics and designed to show the mechanical nature of sound. But because of the small energies in sound waves, mechanical measurement technologies could not match the human ear, which remained the most sensitive sound detector until the advent of electric amplification. Indeed, the very success of acoustics research within the program of nineteenth-century physics actually led to a loss of interest in the field by the turn of the century: physicists thought that acoustics had been sufficiently and systematically researched and did not expect any more fundamental breakthroughs. As physicists lost interest, however, musicologists and psychologists continued to discuss Helmholtz's acoustics. *On the Sensations of Tone* prompted an intense debate about the nature of combination tones and beats. That debate was not resolved but shifted from the objective physical existence of these effects to their subjective psychological perception (Pantalony 2009, 168–170).

While discourses in acoustics moved toward the vibrant fields of experimental psychology and ethnomusicology, physicists turned to electrodynamics as the scientific frontier. The rise of electrodynamics led to the fusion of acoustics with electric technology and electroanalog thinking. Heinrich Hertz's experiments on the propagation of electric waves in 1887 and 1888 were a breakthrough for Maxwell's electromagnetic theory of light (Darrigol 2000, 252–264). At the same time as electromagnetic theory was developing apace, physicists, engineers, and entrepreneurs set out to conceive systems of wireless telegraphy based on Hertz's electric waves, which were generated by alternating currents and propagated in free space (Hong 2001, 1–23). In Victorian Britain, Maxwell and others developed a field theory of electrodynamics that mobilized a mechanical understanding of the configuration of the ether as powerful illustrations of electric phenomena (Buchwald 2013, 572–578). Analog thinking, the transfer of mathematical formalism, and the use of electroacoustic technologies all brought acoustic and electrical research together. Helmholtz had already used electromagnetically driven tuning forks in his experiments for *On the Sensations of Tone*, and the Bell telephone became a popular detector of small alternating currents and their frequency, which could simply be listened to. Telegraph operators, too, started to listen to the telegraph signals through headphones. The notion of electroacoustics itself, combining electric technology and electromagnetic thinking with acoustics, finally appeared around 1900 (Hartmann-Kempf 1903).

The singing or speaking electric arc promised great potential as an electroacoustic technology in the first decade of the twentieth century. After William Du Bois Duddell succeeded in building a circuit for a continuous arc that produced undamped electric waves, it became possible to use arc transmitters for wireless telephony as a way of broadcasting sound rather than telegraph signals. Within different circuit arrangements, the arc could be made to act not only as a transmitter or detector of electromagnetic waves, but also as a loudspeaker or microphone. In Germany, Hermann Theodor Simon at the University of Göttingen developed the electric arc as a versatile experimental system and embedded it in the electromagnetic worldview of the fin de siècle (Simon 1911). Nonetheless, neither the electromagnetic worldview nor the electric arc lived up to their promise. During World War I, it was not the electric arc but the vacuum tube that came to fulfill these functions as a multipurpose circuit component.

2.2 Helmholtz's *On the Sensations of Tone* and Rayleigh's *The Theory of Sound*

Helmholtz's *On the Sensations of Tone* and Rayleigh's *The Theory of Sound* were simultaneously very similar and remarkably different in their scientific aims, in their representations of sound phenomena, and in the readerships they addressed. Without being able to discuss the two works in depth here, I will point out some aspects that are important to the further discussion.

In *Sensations of Tone*, Helmholtz tried to achieve a synthesis of science and music by grounding the human sensations of musical sounds in the physics of sound vibrations and the physiology of hearing. Physics and physiology were to serve as a scientific foundation on which the aesthetic perception of music and its theoretical structure would rest. Helmholtz's involvement with acoustics was, like most of his work in the Heidelberg period, informed by his approach to physiology—the reduction of all natural phenomena to the simple laws of mechanics. While both his optics and his acoustics treated issues of sensation and perception, Helmholtz's acoustics went further than his optics in aiming to place the aesthetic perception of music on a scientific basis.[5] Physics and physiology were to be reduced to mechanics, but aesthetics certainly was not. Reaching out beyond physicists and physiologists, Helmholtz sought out musicologists and aestheticians as the readership of *Sensations of Tone*, commenting that "the horizons of physics, philosophy and art have of late been too widely separated" (Helmholtz 1863, 1). One might argue that precisely for acoustics, the connection between science and art had not actually been broken as Helmholtz claimed. The acoustic work of Ernst Chladni and Wilhelm Weber, for example, was closely related to the understanding and improvement of musical instruments (Jackson 2006). But Helmholtz wanted to go beyond explaining and improving musical instruments. His aim was nothing less than to base the aesthetics of music itself on the physics of sound and the physiology of tone sensation.

Helmholtz had discussed the relationship between the sciences and the humanities in a commemorative address to the University of Heidelberg in 1862. The sciences and humanities shared, he proposed, the same "purpose of making the spirit rule over the world. While the humanities work directly on … separating the pure from the impure, the sciences follow the same objective indirectly by seeking to liberate the human being from the necessities that intrude on him from the external world" (Helmholtz 1896, 183).[6] These words reveal a core aspect of Helmholtz's idealist epistemology of the sciences. As Michael Heidelberger (1994, 168–175) has argued, Helmholtz's epistemology was deeply rooted in Johann Gottlieb Fichte's idealist philosophy.[7] The free human spirit had to liberate itself from the forces of nature by subjecting nature to laws. This almost inevitably created a tension between Helmholtz's reductionist program of physiology and the free aesthetic spirit, a tension addressed in the preface to the third edition of *Sensations of Tone* in 1870. There, he discussed the critical reception that his theory of music had received in published reviews: some of Helmholtz's critics argued that mechanical explanations should be extended to cover artistic and aesthetic aspects, while others rejected the reductionist program of tone sensation as too mechanical. From the nature of criticism from both sides, Helmholtz concluded that he had followed the right path (Helmholtz 1870, ix).

In Helmholtz's view, there was no justification for the discipline of acoustics as a subdiscipline of physics without the sensation of human hearing, since

this *physical acoustics* is essentially nothing but a section of the theory of the motions of elastic bodies. ... In physical acoustics, therefore, the phenomena of hearing are taken into consideration solely because the ear is the most convenient and handy means of observing the more rapid elastic vibrations, and the physicist is compelled to study the peculiarities of the natural instrument which he is employing, in order to control the correctness of its indication. (Helmholtz [1863] 1875, 4)[8]

The direct interdependence of the physical discipline of acoustics and the human sensation of hearing limited the scope of acoustics to those mechanical vibrations that could be perceived by the human ear.

Helmholtz narrowed the scope of his investigations even further. He did not study the myriad sounds that reached the human ear every day, but focused only on *musical* sounds. There seemed to be no difficulty in making the distinction. In Helmholtz's understanding, music was made up of harmonious tones; we look in vain for the notion of "noise as nuisance" (Lärm or Störschalle) in his work. In contrast to this, the distinction between irregular sounds (Geräusche) and the harmonious sounds of music (musikalische Klänge) is central to Helmholtz. In the section on "Noise and Musical Tone" (Geräusch und Klang), Helmholtz divides the world of sound accordingly:

The first and principal difference between various sounds experienced by our ear, is that between *noises* [Geräusche] and *musical tones* [musikalische Klänge]. The soughing, howling, and whistling of the wind, the splashing of water, the rolling and rumbling of carriages, are examples of the first kind, and the tones of all musical instruments of the second. Noises and musical tones may certainly intermingle in very various degrees, and pass insensibly into one another, but their extremes are widely separated.

The nature of the difference between musical tones and noises, can generally be determined by attentive aural observation without artificial assistance. We perceive that generally, a noise is accompanied by a rapid alternation of different kinds of sensations of sound. Think, for example, of the rattling of a carriage over granite paving stones, the splashing or seething of a waterfall or of the waves of the sea, the rustling of leaves in a wood. In all these cases we have rapid, irregular, but distinctly perceptible alternations of various kinds of sounds, which crop up fitfully. ... On the other hand, a musical tone strikes the ear as a perfectly undisturbed, uniform sound which remains unaltered as long as it exists, and it presents no alternation of various kinds of constituents. To this then corresponds a simple, regular kind of sensation, whereas in a noise many various sensations of musical tone are irregularly mixed up and as it were tumbled about in confusion. We can easily compound noises out of musical tones, as, for example, by striking all the keys contained in one or two octaves of a pianoforte at once. This shews us that musical tones are the simpler and more regular elements of the sensations of hearing, and that we have consequently first to study the laws and peculiarities of this class of sensations.

Then comes the question: On what difference in the external means of excitement does the difference between noise and musical tone depend? ...

Our definition of periodic motion then enables us to answer the question proposed as follows:— *The sensation of a musical tone is due to a rapid periodic motion of the sonorous body; the sensation of a noise to non-periodic motions.* (Helmholtz [1863] 1875, 11–12)

In this remarkable account, Helmholtz conflates the physical definition of noise as a nonperiodic motion with the aesthetic notion of noise as the antithesis of a musical sound. Interestingly, Helmholtz takes not the physical but the musical definition as a point of departure to build his argument. To Helmholtz and his imagined readership, it seemed obvious that there was an essential difference in how we perceive musical tones on the one hand and noises on the other. Helmholtz's framing of noise as Geräusch, between physics, sound perception, and aesthetics, continued to characterize the discourse of noise until the interwar period.

In the remaining 582 pages of *Sensations of Tone*, the concept of noise has no life of its own. Helmholtz comes back to noises a few times when they become important for his musical agenda. He points out that musical tones are accompanied by noises. Far from being nuisances, these noises, such as the scratching and rubbing of the violin bow and the noise of rushing winds in flutes and organ pipes, characterize the instruments and make their sound distinctive and interesting to the human ear (Helmholtz [1863] 1875, 101, 114). As for the human voice, in Helmholtz's account it is relevant only for singing and not for speaking. Helmholtz divides the human voice, like all sounds, into musical tones, which he identifies as vowels, and noises, which he identifies as consonants. Again, after making this distinction in the first section of the book, he does not find it necessary to come back to consonants at a later stage (28); as he does with musical instruments, he discusses the noises that accompany vowels (111).

In Helmholtz's work, the perception of sounds, and the categorization of whether a sound was a noise or a tone, was always carried out by Helmholtz's trained ear, aided by instruments such as the tuning fork and the resonators he had developed himself. First, in his view, no measurement instrument was more sensitive than the human ear; second, music always remained the reference point, so he saw no reason to move away from the trained musical ear. The acoustics investigator, in fact, required the refined hearing and touch of a musical performer. For Helmholtz, the theory of sound was both a product of and constitutive of how he, as an individual, heard and created sound (Hui 2013, 57, 60). His own conscious listening constituted not only his self-understanding as a musical listener, but also an experimental epistemology which he presented in the third volume of his handbook of physiological optics:

> By comparatively few carefully executed experiments I am enabled to establish the causal conditions of an event with more certainty than can be done by a million observations where I have not been able to vary the conditions as I please. ... The same great importance which experiment has for the certainty of our scientific convictions, it has also for the unconscious inductions of the perceptions of our senses. It is only by bringing our organs of sense in various relations to the objects through our own will [nach eigenem Willen] that we learn to judge with certainty about the causes of our sensations, and this kind of experimentation begins in earliest youth and continues all through life without interruption. (Helmholtz [1867] 1925, 30–31)[9]

Helmholtz's scientific program of reducing all physical and physiological phenomena to the laws of mechanics stood in contrast to his epistemology of observation and experimentation as matter of free will. The natural world, including the human sensory apparatus, could be described using the laws of mechanics. Yet Helmholtz's observation was not mechanical observation. In Helmholtz's epistemology, nature had to follow the mechanical laws of causality; our own will, in contrast, was free, and only through our free will could we learn about nature and determine its laws. Helmholtz admitted that many people were unable to repeat his experiments in separating combination tones or upper harmonics from fundamental tones, even when using his instruments. Rather than questioning his own observations, however, he attributed this apparent failure to a lack of mental effort and attention by the individuals concerned (Helmholtz 1867, 432).

If Helmholtz wanted to bring together science, art, and philosophy, Rayleigh's goal was entirely different: he aimed to lay "before the reader a connected exposition of the theory of sound, which should include the more important of the advances made in modern times by Mathematicians and Physicists" (Rayleigh 1877, v). Helmholtz had restricted his use of higher mathematics, especially differential calculus, to make his volume accessible to readers from musicology or physiology who did not necessarily have special training in mathematics or physics. Rayleigh's book, in contrast, is filled with differential calculus—it is a handbook that provides the trained reader with the tools for a mathematical treatment of all kinds of acoustic phenomena. *The Theory of Sound* was a model of how mathematical physics in the Cambridge tradition should proceed: posing a physical problem and subjecting it to rigorous mathematical analysis. The approach to the problem, its framing, and the mathematical solution could then be applied to similar physical problems. These similar cases might be found in acoustics, but also in entirely different fields. The mathematical treatment of acoustics provided an excellent training for the mathematical treatment of all kinds of oscillation and vibration problems. From this perspective, Rayleigh's paragraphs in *The Theory of Sound* read like a list of problems set for students in an exam in mathematical physics. By analogy, differential equations and their solutions derived from acoustic problems could be applied in optics, electrodynamics, and thermodynamics. Rayleigh's rigorous treatment of acoustic wave mechanics and wave propagation made *The Theory of Sound* particularly useful as a vast reservoir of analogies for the mathematical study of electromagnetic wave propagation. In the twentieth century, physicists adopted Rayleigh's methods and solutions when approaching problems in general relativity and quantum mechanics.[10]

These differences should not obscure the similarities between Helmholtz's and Rayleigh's approaches to acoustics. Both treated sound as an essentially mechanical phenomenon, which was explained by oscillations in solids, liquids, and gases. Both agreed that complex sounds were assembled through the combination of simple tones (sine waves), and that these sounds could be analyzed as such by applying the Fourier theorem. Helmholtz's argument that acoustic oscillations were limited to audible frequencies alone also allowed Rayleigh to keep acoustics, as a field of inquiry, separate from mechanical vibration. Though Rayleigh wrote for physicists rather than musicologists, he nevertheless followed Helmholtz's regime of separating the world of sound into musical sounds and noises. Early in the first volume of *The Theory of Sound*, he explains:

> Before proceeding further we must consider a distinction, which is of great importance, though not free from difficulty. Sounds may be classed as musical and unmusical; the former for convenience may be called *notes* and the latter *noises*. The extreme cases will raise no dispute; every one recognises the difference between the note of a pianoforte and the creaking of a shoe. But it is not so easy to draw the line of separation. ... [A]lthough noises are sometimes not entirely unmusical, and notes are usually not quite free from noise, there is no difficulty in recognising which of the two is the simpler phenomenon. ... [B]y sounding together a variety of notes—for example, by striking simultaneously a number of consecutive keys on a pianofortes—we obtain an approximation to a noise; while no combination of noises could ever blend into a musical note. (Rayleigh 1877, 4)

After this short passage in the introduction, Rayleigh never again mentions noise in his otherwise rather comprehensive two-volume treatise on acoustics. Given that Rayleigh claimed to "lay before the reader a connected exposition of the theory of sound, which should include the more important of the advances made in modern times by Mathematicians and Physicists" (Rayleigh 1877, v), this lack of further discussion implied that scientists were not interested in noise. Not only for Helmholtz but also for his contemporary acousticians, sound was defined as musical sound, and noise—as the opposite of music—seemed to have no space in this world.

The restriction of acoustics to musical tones meant that acousticians only looked at a small fraction of the soundscape. This understanding of sound as music reflects the central role of music as high culture in the bourgeois academic communities of the late nineteenth century, which idealized sound as (classical) music, and musical listening as the only form of listening. Harmonic sounds were also the kind of sounds that acousticians could deal with. Musical harmonic tones were resonant tones that could be made visible. It was, and still is, one of the most powerful methods of mathematical and experimental physics to reduce complex phenomena to simple representations. Helmholtz's and Rayleigh's acoustics took simple tones as a point of departure, or separated a more complex tone into its simple harmonics. Helmholtz and Rayleigh would of course argue that every complex sound could be assembled from, or deconstructed into, a series of simple tones of different frequencies and amplitudes by means of the Fourier theorem; simple tones could be analyzed with the mathematical apparatus of differential calculus as well as with experimental apparatus such as tuning forks and Helmholtz resonators. On the other hand, it meant that Helmholtz and Rayleigh simply did not engage with complex sounds or noises.

The tuning fork embodied Helmholtz's theories and his understanding of acoustics (Pantalony 2009, 22–24). It was the most accurate acoustic instrument, assuring precision in the laboratory as well as in the concert hall. In the tuning fork synthesizer, Helmholtz combined the simple tones of tuning forks with different pitches to create more complex vowels (Helmholtz 1863, 182–193). Even the vibrations of tuning forks contained upper harmonics, which Helmholtz spent some time investigating. By attaching a stylus to the fork, he made a sounding tuning fork write its own vibration on paper (33–35). Because of its regular vibrations, the tuning fork could be applied as a precision instrument in fields other than acoustics. Helmholtz himself used a tuning fork as a mechanical interrupter for the current in his tuning fork synthesizer (figure 2.1). The graphical method of a tuning fork writing its own vibration on a moving sheet of paper enabled precise time measurement. In 1868, Wilhelm von Beetz introduced an electric tuning fork chronoscope that did not require a clockwork mechanism.[11] Helmholtz also employed Chladni figures to make vibrations on discs visible and observed string vibrations through his vibration microscope (Helmholtz 1863, 33–34, 123, 137–148). Through these visual or graphic representations, Helmholtz could observe not only fundamental tones, but also the upper harmonics and disturbances of musical tones. Helmholtz argued that the method of attaching a pin to draw a graphic representation could be used for all resonating bodies. Sand figures similar to the Chladni figures could be produced on all kinds of resonating bodies, including membranes stretched over open resonators (64–66).

When it came to airborne sound, however, the ear remained the only practical sound detector, aided by the spherical resonators developed by Helmholtz. Using the Helmholtz resonator to listen to its particular resonance tone was only possible while blocking out all other frequencies of a more complex sound (Helmholtz 1863, 73–75). In 1882, Rayleigh introduced a disc suspended by a torsion thread as an objective instrument

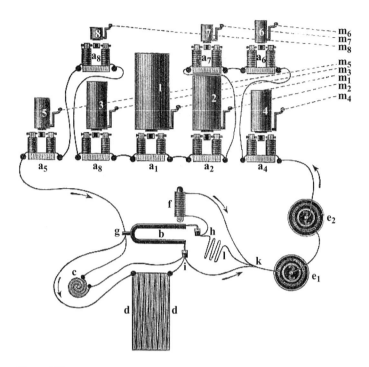

Figure 2.1
Helmholtz's wiring diagram for his electromagnetically driven tuning fork synthesizer (Helmholtz 1863, 584). The diagram follows very different conventions from those of twentieth-century circuit diagrams. Components a_1 to a_8 represent the electromagnets driving the tuning fork in front of the respective resonator 1 to 8; b is the interrupter for the circuit, which itself is driven by the coil f; e_1 and e_2 represent two Grove cells; c is a condenser, and d is a "very great" resistance.

to measure sound intensity for airborne sound. In contrast to Helmholtz's instruments, the Rayleigh disc did not select frequencies but added up the sound pressure over all frequencies. Rayleigh was immediately struck by the differences between measured sound intensities and the intensities that he perceived by listening (Rayleigh 1882). His observation suggested that the human ear could not be trusted as an indicator for sound intensity when compared with measurements using physical methods, which were associated with the concept of objective measurement. The relationship between physical sound pressure and subjectively perceived sound intensity then became the subject of experimental psychology.

2.3   Experimental Psychology and Beyond: From the Physics of Sensation to the Evolution of Music

On the Sensations of Tone was highly influential for the development of physical acoustics, physiological acoustics, and musicology alike; however, Helmholtz's view of music as high culture, and the universality of concepts such as consonance and dissonance, became rather fragile around the turn of the century.[12] Instrument maker Rudolph Koenig, based on his own experiments, challenged Helmholtz's theory of combination tones in the late 1870s. Toward the end of the nineteenth century, the discourse about consonance and combination tones increasingly moved from physics and physiology to the emerging field of experimental psychology. Experimental psychology had its roots in earlier traditions of psychophysics, represented by Gustav Fechner and Ernst Weber in Göttingen, but Helmholtz himself and his work on sensation also played a major role in the formation of the discipline. Wilhelm Wundt, one of the leading founders of experimental psychology, was Helmholtz's assistant during his work on *Sensations of Tone*. Another pioneer of experimental psychology, the experimental philosopher Carl Stumpf, became one of Helmholtz's most eminent critics in the debate (Stumpf 1883; Hui 2013, 123–148).

As much as Stumpf and Koenig disagreed with Helmholtz on combination tones and beats, both acknowledged his important role and his contributions to the scientific foundations of acoustics. While praising him for the stimulation and support he had given to the study of tone psychology, Stumpf argued that Helmholtz had left much undone (Stumpf 1883, v). Around the turn of the century, comparative musicology and an evolutionary understanding of sensual taste and music challenged the idea that aesthetics could be fundamentally based on physics and human physiology. Far from damaging the popularity of Helmholtz's acoustics, however, Koenig's and Stumpf's criticisms underlined its importance through the debates they created.

It would be a mistake to believe that Helmholtz's acoustics spread solely by means of his writings. Listening with Helmholtz did not only mean listening to Helmholtz. It meant using his instruments, repeating his experiments, and observing the sound effects he described. The decades following the publication of *Sensations of Tone* saw an unprecedented expansion of scientific and technical education in Germany and elsewhere. The growth of scientific education embraced primary and secondary schools, as well as technical colleges and universities. Physics was now taught not only by the textbook but through experiments. Although teaching laboratories became common, with students carrying out simple experiments, much if not most experimental physics teaching was done in lecture demonstrations. Wealthier teaching institutions and universities acquired impressive teaching collections of elaborate instruments to demonstrate all kinds of physical phenomena (Brenni 2011). These physical cabinets typically had a section on acoustics, including the instruments introduced by Helmholtz. Historical acoustic demonstration apparatuses are still numerous in collections held by institutions ranging from primary schools to university departments, and many show signs of decades of heavy use. The teaching of acoustics through these instruments created an attractive link between science and music. It also enabled a vivid understanding of vibration and wave propagation that was fundamental to physics beyond the field of acoustics. Some of the instruments, such as the Koenig sound analyzer and the Chladni discs, visualized sound and made vibrations *anschaulich*, in the sense that physics became clear by *looking at* them. With most of the acoustic instruments, however, vibrations became clear by *listening to* them.

Rudolph Koenig was the maker of the most prestigious of these acoustic instruments.[13] Koenig grew up in Königsberg in East Prussia, where Helmholtz had been professor of physiology for a short while. In 1851, Koenig moved to Paris, at that point the mecca of scientific instrument making. He trained as a violin maker and took over the workshop of the acoustic instrument maker Albert Marloye in 1858 (Pantalony 2009, 9). Koenig worked with Édouard-Léon Scott on the phonautograph. Like the tuning fork, the phonautograph could record vibrations graphically. Sound vibrations were directed onto a membrane by a funnel and written down by a stylus connected to the membrane. The phonautograph was similar to the phonograph in that it produced a curve as a graphic record of a sound on a cylinder, but it could not reproduce the sound. From 1859 onward, Koenig corresponded with Helmholtz and started to manufacture Helmholtz's instruments (51–52). Koenig did not simply build Helmholtz's instruments; he also carried out experiments and observations, developed his own instruments, and improved the designs of Helmholtz and others. Koenig's most important contribution to the array of acoustic instruments was the manometric flame capsule, which he used to visualize sound vibrations by projecting rhythmically flickering gas flames onto a rotating mirror (88–92).

Based on his own experiments and observations, Koenig disagreed with Helmholtz's description and interpretation of combination tones. Helmholtz had rejected the idea that combination tones had only subjective existence, originating from the nervous system. Instead, combination tones corresponded to *real* vibrations in the eardrum and auditory ossicles (Helmholtz 1856, 537). Helmholtz traced combination tones that had physical origin produced by musical instruments but concluded that only the resonating parts of the ear produced the strong combined oscillations that led to the sensation of combination tones (Helmholtz 1863, 235–236).

Koenig later claimed that he had originally set out to confirm Helmholtz's theory of combination tones, but after accumulating a large set of observations, he came up with his own concept of the *Stoßton* (commonly translated as "beat note").[14] Koenig's results seemed to agree with Thomas Young's older beat theory, which Helmholtz had opposed (Pantalony 2009, 96–100; Koenig 1876). Koenig's precision tuning forks were key instruments in his experiments, observations, and arguments. Not having an academic affiliation, he put into play his reputation as an instrument maker and demonstrations of his instruments to visiting scientists in his Paris workshop. Outside the workshop, Koenig's instruments became very successful for demonstrations and teaching in experimental lectures. Koenig succeeded on the educational market but otherwise found it difficult to sell his highly specialized and precise research apparatus (Pantalony 2009, 129–130).

Whereas Koenig's critique of Helmholtz's combination tones came from his practice of instrument making, Stumpf's critique came from philosophy, musicology, and musical listening. Stumpf had studied philosophy with Franz Brentano in Würzburg and Hermann Lotze in Göttingen. In Göttingen, Stumpf came into contact with Ernst Weber and Gustav Fechner, the founders and most prominent representatives of psychophysics. In 1873, Stumpf returned to Würzburg, where he worked on aesthetics and perception as professor of psychology. In 1875, he started work on *Tonpsychologie* (Psychology of tone), one of his major contributions to psychology, which was published in two volumes in 1883 and 1890 (Sprung and Sprung 2006, 203–240). For Helmholtz, combination tones originated in the sensory apparatus and were *objective*, physical, and physiological in nature. For Stumpf, they were *subjective* and psychological in nature. He contrasted the Helmholtzian concept of *Tonempfindung* (sensation of tone) with the concepts of *Tonwahrnehmung* (perception of tone), *Tongefühl* (feeling generated by tone), *Tonvorstellung* (mental representation of tone), and *Tonurteil* (tonal judgment).

The phonograph, introduced by Thomas Edison in 1877, offered radically new perspectives for the study of sound and music. For the first time, the vibrations of airborne sounds could be mechanically recorded and reproduced. The phonograph and Emil Berliner's invention of the gramophone made it possible to record, store, and later replay speech and music. Linguists and musicologists started to collect and archive sound portraits of famous people, samples of spoken language, and musical performances. Stumpf, who had been appointed professor of philosophy at the University of Berlin in 1894, initiated the Berlin Phonogram Archive in 1900, one year after the Austrian Imperial Academy of Sciences founded its own Phonogram Archive (Stangl 2000, 121–148). Stumpf appointed his student Erich Moritz von Hornbostel as the first director of the Berlin Phonogram Archive. Hornbostel had studied chemistry with Adolf Lieben in Vienna before he became Stumpf's assistant in Berlin in 1900. The archive collected phonogram recordings of folk music from different cultures around the world (Stumpf 1908; Hui 2013, 139–144).

Hornbostel regarded Alexander Ellis, the English translator of Helmholtz's *Sensations of Tone*, as the father of comparative musicology. In Ellis's comments to the translation, he compared the German vowel and consonant sounds to those in other languages, among these various English dialects, Sanskrit, and Bengalese (Helmholtz [1863] 1875, 107–108, 727–741). In 1885 he published a comparative study on the intervals of musical instruments across the world (Ellis 1885). Hornbostel, however, found Ellis's and other studies of musical instruments alone as faulty for determining a musical system, as the tonality of the instruments depended crucially on the practice of the music performer (Hornbostel 1905, 89–90).

The phonograph facilitated objective recordings of music, similar to the objectivity of photography, which could then be transported, duplicated, and preserved for an indefinite period. Earlier methods of writing down "exotic" (that is, non-European) music in European classical music notation became highly problematic when compared to the phonograph recording. According to Stumpf, European musical notation and the whole theoretical system of European classical music had led to a wrongful Europeanization in the understanding of non-European music. The music of people seen by the Europeans as primitive (usually called *Naturvölker* in German), but also the music of the civilized Chinese, Indians, and Persians, had been made to fit a European (and, we would have to add, a high-culture) understanding of music and its development.[15] By separating the music from the performer, the phonograph enabled new means of disembodiment and a scientifically desirable distance between the observer and the observed. The phonograph recordings allowed researchers to study rhythmic and tonal peculiarities with the scientific precision of the tonometer and metronome for the first time, transforming comparative musicology into a scientific discipline. The practice of musical listening as high culture, until then seen as a necessary scientific skill, was now only damaging to this scientific approach.

While Helmholtz had located the foundations of music in timeless physical and physiological principles, Stumpf, Hornbostel, and other comparative musicologists were searching for its historical or even prehistorical origins (Stumpf 1911; Rehding 2000). Helmholtz did not want to reduce music to the simple laws of mechanics either. He distinguished between sensuous pleasure, which belonged to science, and aesthetic beauty, which belonged to the discipline of aesthetics. Although the first, scientific principle was "an important means for attaining the second," aesthetic principle, they had to be kept strictly apart (Helmholtz [1863] 1875, 357–358). In chapter 13 of *Sensations of Tone*, he laid out his aesthetic method, which was essentially historical. Following classical musicology, he searched for the origins of music in antiquity. He identified this music as homophonic, medieval music as polyphonic, and modern music as harmonic. Chinese, Indians, Arabs, Turks, and modern Greeks remained, Helmholtz argued, caught in the stage of homophonic music (361). Helmholtz's history of music largely followed the Europeanizing approach to music, which Stumpf later criticized. "[U]ncultivated or savage people" (358) did not enter Helmholtz's aesthetic argument in any way.

In 1871, Charles Darwin brought a new scientific principle to the debate on the origins of music, the principle of evolution (Ames 2003, 302). In *The Descent of Man*, Darwin located the origins of music not in ancient Greece but in the animal kingdom. The evolutionary narrative of the origins of music opened up a whole new perspective for ethnographers, who attributed a lower evolutionary level to those people they saw as primitive and savage. If the music of exotic peoples was regarded as primitive, then it could parallel earlier stages of European music, as Hornbostel and his colleague Otto Abraham argued (Abraham and Hornbostel 1904, 224–225; Hornbostel 1905, 96; Ames 2003).

Like other phonogram archives of ethnomusicology, the Berlin Phonogram Archive was part of the colonial scientific agenda, directed at dominating and exploiting the colonies not only economically but also scientifically (Stumpf 1908, 245–246). Stumpf and Hornbostel expressed concern that the languages and music of native peoples were vanishing because of the spread of European culture and the disappearance of tribes and tribal structures (Stumpf 1908, 239–240; Hornbostel 1905, 97). Nonetheless, while the ethnographers were preserving the music and language of people they perceived as primitive, the colonizers were marginalizing those very people and pursuing their extinction. In 1905, Hornbostel associated the method of comparative musicology with comparative anatomy. Just as comparative anatomy gave clues about the classification and development of organs across species, comparative musicology could give clues about the evolutionary development of music across cultures. Hornbostel propounded the value of "atomizing" [sic] the music of the Herero and other Khoikhoi tribes of the colony of German South West Africa, with scientific precision by using the tonometer and metronome (Hornbostel 1905, 85; Ames 2003).

While questioning the artistic value of the music of the Hottentots, as the Herero were called by the colonizers, Hornbostel emphasized its scientific value for studying the origins and development of music. If we think of ethnomusicology as colonial science, Hornbostel was at war in 1905. Between 1904 and 1908, the German Empire fought the rebellious Herero and Nama tribes and committed a genocide that killed about 100,000 Africans, or about 80 to 85 percent of the Herero population. The classification of the Herero as uncultivated and primitive by Hornbostel and other ethnographers and anthropologists formed part of the undertaking to devalue their lives and humanity. It effectively served as a justification for the atrocities committed by the colonizers against the colonized (Zimmerer and Zeller 2004).

Ethnomusicology was not the only threat to Helmholtz's understanding of music and his division of sounds into musical tones and noises. With the growth of experimental psychology, the aesthetic values of psychophysics lost their appeal for scientists (Hui 2013, 46). In the years before World War I, Italian futurists launched a full-fledged attack on the aesthetics of musical sounds. The poet Filippo Tommaso Marinetti founded futurism as a vanguard revolutionary art movement. In his 1909 manifesto "Le Futurisme," Marinetti glorified violence, warfare, modern technology, and urban industrial life, calling for the destruction of museums, libraries, and academies as representatives of tradition. Luigi Russolo's manifesto *L'arte dei Rumori* (*The Art of Noises*) of 1913 embraced the roar of the modern city and industrial life, attacking the bourgeois order of music and noise.[16]

The futurist movement was fascinated by science and its newest achievements. But science was not an end in itself. Scientific concepts and methods were means to achieve an aesthetic vision, not to produce science. Luciano Chessa has argued that the futurist understanding of science was one of scientific occultism rather than scientific positivism (Chessa 2012, 16–17). Russolo was not trained in music, and he worked as a visual artist prior to *L'arte dei Rumori*. He had studied wave theory and the works of Chladni and Helmholtz, whom he cited, and he had carried out experiments with X-rays (Chessa 2012, 119–120). Going beyond physics, his artistic project introduced evolution as a principle for change in society. Russolo adopted the idea of musical evolution from the ethnomusicologists and placed it at the heart of his argument that music must necessarily undergo change in an industrialized and urban world. Music would evolve from harmony and consonance to dissonance and stranger, harsher sounds. This evolution from Beethoven and Wagner to the noises of trams and backfiring motors, Russolo argued, only became possible as an adaptation to the machine age. In his understanding, musical evolution paralleled the proliferation of machines just as the roar of machines created a sound rivaling pure sounds, which in the new surroundings appeared monotonous and unable to provoke any emotions. In Russolo's manifesto, music had "searched out the most complex succession of dissonant chords, which have prepared in a vague way for the creation of **MUSICAL NOISE**" (Russolo [1913] 1986, 24; original emphasis):

We want to give pitches to these diverse noises, regulating them harmonically and rhythmically. Giving pitch to noises does not mean depriving them of all irregular movements and vibrations of time and intensity but rather assigning a degree or pitch to the strongest and most prominent of these vibrations. Noise differs from sound, in fact, only to the extent that the vibrations that produce it are confused and irregular. *Every noise has a pitch, some even a chord, which predominates among the whole of its irregular vibrations.* Now, from this predominant characteristic pitch derives the practical possibility of assigning pitches to the noise as a whole. That is, there may be imparted to a given noise not only a single pitch but even a variety of pitches without sacrificing its character, by which I mean the timbre that distinguishes it. Thus, some noises obtained through a rotary motion can offer an entire chromatic scale ascending or descending, if the speed of the motion is increased or decreased. (27; original emphasis)[17]

Russolo divided noises into six categories that the futurist orchestra would set into motion mechanically. The categories included the noises of guns and machinery. Between 1912 and 1914, Marinetti published his sound poem "Zang Tumb Tumb," which describes the sound of gunfire and explosions from the trenches of Adrianopolis (Edirne) in the First Balkan War. Russolo cited Marinetti's description of the firing of guns as an orchestra, which embraced the sound of war and effectively predicted the "symphony of the industrial war." Acousticians actually started to listen to and analyze gun and machine noise in World War I, a year after the publication of *L'arte dei Rumori.* We will come to this story in the next chapter. Russolo's manifesto remained silent, however, on the most important revolution (or, with Russolo, we might argue for evolution) in sound technology and its industrialization then under way: the production and rendering of sound by electrical technologies, and the fusion of acoustic with electroanalog thinking.

2.4 Electrification of Sound: Electrodynamic Theory, Instruments, and Circuit Design

The most important change occurring in acoustics around the turn of the century was beyond doubt its transformation into electroacoustics. The notion of electroacoustics referred to electroanalog technologies such as the telephone, which transformed acoustic vibrations into electric oscillations and vice versa. But electroacoustics stood for more than simply a technological change. It also signified the use of electroacoustic analogies as a language that allowed physical problems and their mathematical formalism to be translated from acoustics into an equivalent problem in electromagnetism.

To understand the emergence of electroacoustics as a concept around 1900, it is important to look at the development of electromagnetic theory as well. In the mid-nineteenth century, electric phenomena were not yet understood as oscillations or waves. Physicists were guided by the paradigm of a unified mechanical worldview and tried to reduce electricity and magnetism to the simple laws of mechanics. Heinrich Hertz's experiments on the propagation of electric waves in free space finally turned the understanding of oscillatory electromagnetic phenomena upside down and brought them to the forefront of scientific research (Darrigol 2000, 252–262). Mechanical models and the analogy with acoustic vibrations and wave propagation became useful to make sense of these electric oscillations and waves.

Both Helmholtz and Rayleigh drew on emerging electric technologies and electrical analogies in their work on acoustics. Helmholtz used telegraph technology in his acoustic investigations, employing electromagnets and interrupters to drive the tuning forks for his tuning fork synthesizer (see figure 2.1) and his vibration microscope (Helmholtz 1863, 137–139). He also employed telegraph wires as an analogy, explaining that the nerves convey sensation in the same way that wires convey electric telegraph pulses (222–223). Helmholtz's representations of sound were visual, however, not electrical or electroanalog. He did not deploy an analogy between electric oscillations and acoustic vibrations, but the analogy between perceptions of sound and perceptions of vision was important. This reflects the understanding of both electrical oscillation and communication technology at the time. Telegraph signals were understood as pulses, which could drive Helmholtz's tuning forks, and not as electric waves, which could be translated into equivalent acoustic waves by means of transducers.

The physicists' comprehension of electric oscillations changed rapidly with the advance of electrodynamic theory, the design of electric circuits, and the development of the telephone as a reversible electric sound transducer. Helmholtz himself had an important influence, if not an active role, in all three of these developments. Helmholtz's interest in electricity was, like his work in acoustics, initially informed by his research in physiology. In the 1850s he conducted a number of experiments in which he measured the propagation velocity of nervous excitation in frog legs caused by electric stimulation. His experiments on nervous excitation led him to the quantitative treatment of self-induction and the extension of Ohm's law to variable currents (Darrigol 2000, 220). He also determined the voltage-source equivalent, a foundation for the equivalent circuit concept, while working on the measurement of voltages and currents in muscle tissue (Helmholtz 1853; D. H. Johnson 2003a, 2003b). Helmholtz was especially interested in open circuits and oscillatory discharge. In the 1870s, he came back to the question of open circuits when he compared the assumptions and theories of Wilhelm Weber and Franz Neumann with those of Michael Faraday and James Clerk Maxwell in a series of papers on the theory of electrodynamics (Darrigol 2000, 223–233).

Helmholtz's adoption of telegraph technology and the telegraph analogy in his physiological investigations comes as no great surprise. He was a close acquaintance of the industrialist Werner Siemens, who presented his pioneering work on the electromagnetic telegraph at the Berlin Physical Society around 1850, the same period in which Helmholtz presented his work on nerve transmission (Lenoir 1994, 187–188; C. Hoffmann 2003). In fact, in the second half of the nineteenth century, not Prussia but Britain was the undisputed leader in establishing a worldwide network of cable telegraphy, which became the "nervous system" of the British Empire (B. Hunt 1991, 54). Enlarging and improving this telegraph system created a great demand for electrical knowledge. William Thomson, professor of natural philosophy at the University of Glasgow, became chief adviser to the first transatlantic telegraph cable project in 1857 and developed electrical measurement instruments for the telegraph industry. The testing laboratories of the cable industry, rather than university laboratories, were the most sophisticated and best equipped electrical laboratories of the day. The requirements of cable telegraphy led to the development of standard electrical units, precision measurement instruments, electric circuit design, and electrodynamic theory (B. Hunt 1991, 56).

In *A Treatise on Electricity and Magnetism* (1873), James Clerk Maxwell acknowledged the importance of telegraphy to understanding electrodynamics (B. Hunt 1991, 56; Maxwell 1873, 1:viii). In his electromagnetic theory, Maxwell applied analogies not between electric oscillations and acoustics, but between the propagation of electric and magnetic forces in the ether and the propagation of light. This was to support his theory that the electromagnetic ether and the light ether were the same medium, and that light was an electromagnetic disturbance in this medium. Maxwell's electromagnetic theory was basically a hydromechanical theory of this ether. The fluids and gases in which sound waves propagated were different from the electromagnetic and light ether, but the theory of hydromechanics established a link between the imponderable ether and the ponderable media where sound propagated. Maxwell made use of mechanical analogies and models as vivid representations of electrical phenomena, declaring that illustrative mechanical models provided the best way to translate complex mechanical relationships into a concrete and readily grasped form without a loss of rigor (B. Hunt 1991, 75). After Maxwell's death in 1879, a group of Victorian physicists developed his theory further. Firm believers in Maxwell's views, they referred to themselves as Maxwellians. Among these, Francis FitzGerald and Oliver Lodge especially adopted and expanded the use of mechanical models to explain the mode of operation of electromagnetic phenomena within the ether (figure 2.2.). According to FitzGerald, these models were only analogies, or likenesses,

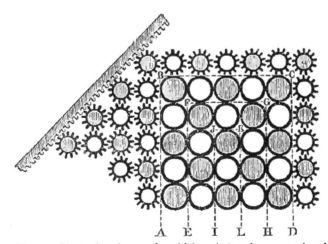

FIG. 43.—Diagram illustrating the way in which an induced current arises in a mass of metal immersed in an increasing magnetic field ; also how it decays. The dotted lines A B C D, E F G H, I J K L, are successive lines of slip.

Figure 2.2
Model illustrating a current induced in metal by an increasing magnetic field (Lodge 1889, 194). Some models of electricity and magnetism, such as Lodge's hydrostatic model of the Leyden jar, were actually built as lecture demonstration devices (see Max Kohl AG 1911, 3:836, fig. 60,638).

and not true representations of how the ether ought best to be imagined (B. Hunt 1991, 83).

Rayleigh's approach to mathematical physics in *The Theory of Sound* became an important resource for Victorian physicists to understand and develop Maxwellian electrodynamics. The analogy between acoustic and electric wave propagation was intuitive for a community that understood electrodynamic phenomena as mechanical phenomena. Even though sound propagation in matter differed fundamentally from electric wave propagation in the ether, the underlying mathematical formalism was analogous and could be transferred from one type of wave to the other. The analogy between sound waves and electromagnetic waves became important when the Maxwellians discussed the electromagnetic production of light. The British concern with what was later coined wave guidance prevented them against thinking that an open oscillating circuit would radiate a substantial amount of electromagnetic energy (Buchwald 2013, 580–581). By 1880, Francis FitzGerald had come to the conclusion that the electromagnetic production of light was impossible, but after carefully studying Rayleigh's work, he had to reconsider this viewpoint. Rayleigh's solution of the equation that FitzGerald was addressing, which took the same form for acoustics as for electromagnetism, suggested that "a simple periodic current would originate wave disturbances such as light" (B. Hunt 1991, 38–42).

It was not the British Maxwellians but Helmholtz's former student Heinrich Hertz who was the first to produce electric waves that propagated in free space, and to establish their affinity to light waves, in 1887 and 1888 respectively. Hertz was looking for interactions between circuits at short distances, not electric waves, when he began his experiments. His point of departure was a prize question posed by Helmholtz in 1879 on the interaction between rapid electric oscillations and dielectric polarization (Buchwald 2013, 578–583; Wittje 2000). Hertz used the analogy between electric oscillations and acoustics to describe the electric effects that he had produced and detected. He was aware that the rapid oscillations in his primary circuit were not particularly harmonic, "but rather such as are produced by striking a wooden rod with a hammer, —oscillations which rapidly die away, and which are mingled [with] irregular disturbances" (Hertz 1893, 49). After he became convinced that his device produced electric waves traveling in free space, Hertz drew the analogy between acoustic and electric waves to argue for the wave nature of the electromagnetic effect that he had created (135–136). Once again, however, it was the analogy with light waves, not acoustic waves, that Hertz had to establish to identify his electric waves as light of very long wavelengths.

Hertz's experiments led to debates about the behavior of his oscillating circuits and their multiple resonances. Their theoretical interpretation introduced Maxwellian field theory to the German scientific community and gave a boost to the further study of electrodynamic theory. Equally important for the development of the electroacoustic agenda was telephony. Alexander Graham Bell's reading of Helmholtz's *Sensations of Tone* may not have accorded with Helmholtz's scientific understanding of the subject, but, as Jonathan Sterne has argued, there was a direct link between Helmholtz's research on the physiology of hearing and Bell's construction of his telephone as a tympanic machine.[18] Interestingly, the telephone started its career in science not as an acoustic instrument, but as an instrument for electrical precision measurement. After Alexander Graham Bell presented his telephone to the public at the Centennial International Exhibition in Philadelphia in 1876, British physicists were among the first to use it as a convenient detector for small oscillating currents. In 1878, George Forbes published a short note in the journal *Nature* reporting that he was able to detect extremely small intermittent currents with the telephone. Whereas the currents were barely perceptible using the galvanometer, they could be easily heard by a croaking or a click in the telephone (Forbes 1878). The telephone even indicated the frequency of intermittence by the tone it produced. In January 1880, Oliver Lodge reported on using a telephone rather than a galvanometer to detect alternating currents in a Wheatstone bridge (Lodge 1880; Rayleigh 1880). The Wheatstone bridge circuit was one of the most convenient and most precise methods for measuring electrical parameters by comparing resistances. The telephone could not easily measure the magnitude of the current— but since the idea of the bridge was to balance the circuit until no current was passing through the telephone, this was not necessary. The British American scientist David Edward Hughes, who had invented a carbon microphone in 1878, used a telephone along with a microphone clock in his induction balance.[19]

Scientists and engineers did not use the telephone only as a measurement instrument; they also tried to improve telephony as a communication system. One of the most important tasks was to increase the range for telephone calls and to decrease distortion as the signals faded away. Oliver Heaviside, a member of the Maxwellians' inner circle and nephew of Charles Wheatstone, quit his job as a telegraph operator in 1874 to pursue his private research. Between 1885 and 1887, Heaviside applied Maxwell's theory to an examination of how electric signals traveled along wires and how their distortion could be reduced or eliminated. Heaviside conceived electric signals in a fundamentally new way, not as pulses in wires, but as trains of electromagnetic waves on wires. Heaviside's discovery of inductive loading in 1887 proved to be extremely valuable for long-distance telephony.[20]

In the first edition of *The Theory of Sound* (Rayleigh 1877, 75), Rayleigh had drawn analogies between acoustic phenomena and electric currents, and introduced the concept of conductivity to acoustics in analogy to electric conductivity (Rayleigh 1878, 159). He did not, however, distinguish analogies between acoustics and electricity or electromagnetism from other analogies, for example, between acoustics and heat conduction, optics, or water waves. When Rayleigh published the second revised edition of his book in 1894, he added a chapter on electrical vibrations. Electrodynamic theory had been transformed by Hertz's experiments and their reception in Britain. Rayleigh had also followed the experiments of Lodge and Hughes attentively (Rayleigh 1880, 1882). To add the theory of electric vibration to a more generalized theory of mechanical vibrations was not a large step for Rayleigh, as "including electrical phenomena under those of ordinary mechanics is exemplified in the early writings of Lord Kelvin; and ... Maxwell gave a systematic exposition of the subject from this point of view" (Rayleigh 1894, §235i). The treatment of electric vibrations in the new chapter, however, is somewhat disconnected from the treatment of acoustic vibrations in the other chapters, and Rayleigh does not use the term "electroacoustics." In section 235, he does discuss the telephone as a transmitter and a receiver of sound, but does not treat it mathematically. A mathematical theory of the telephone, treating both its mechanical and its electromagnetic aspects, was far from trivial and would have been time and space consuming. Henri Poincaré finally presented a comprehensive theory of the telephone in 1907, bringing together Maxwellian electrodynamics of telephony with the mechanical vibration of the membrane in the telephone transmitter and receiver (Poincaré 1907).

The rise and emancipation of electrical engineering was just as important for the advance of electroanalog thinking as electrodynamic theory
and telephony were. Electrical engineers developed standardized circuit
diagrams not only as a design tool but also as a language to represent physical phenomena. Heaviside and Poincaré were masters of Maxwellian electrodynamics. For most electrical engineers and even many experimental
physicists, however, the abstractions of Maxwellian theory remained a
closed book. In the 1880s, electrical engineering was institutionalized as a
new discipline at the German Technische Hochschulen. This process was
driven mainly by the rapid advance of electrification and the heavy-current
electrical industry; telegraphy and telephony played virtually no role.
Electrical engineering emerged as an amalgam of physics and mechanical
engineering, oriented on industrial practice (König 1995). Important as
Maxwell's electrodynamic theory certainly was for understanding electric
oscillations and their propagation both on wires and in free space, it was
too abstract to be taught to engineers and impractical for designing heavy-
current machinery. Electrical engineers needed visual rather than mathematical tools for the design calculations of electrical devices (Kline 1992,
108–112).

One of the pioneers in developing those new tools was Carl Proteus
Steinmetz. He was a leading figure in the theory of heavy-current electrical
engineering and a forceful promoter of alternating current (Kline 1992).
Steinmetz had studied mathematics, physics, and electrical engineering
in Breslau (now Wrocław) and Zurich before he emigrated to the United
States in 1889 and started working in the expanding electrical industry. He
introduced graphical methods and complex numbers into heavy-current
engineering and made ample use of analogies between electricity and
mechanics. In his 1897 textbook on alternating-current phenomena, he
applied equivalent circuit diagrams to explain alternating-current transformers (Steinmetz and Berg 1897, 183–185).[21]

Around the turn of the century, the notion of electroacoustics made
its appearance. In 1903, Robert Hartmann-Kempf added the subtitle
Elektroakustische Untersuchungen (Electroacoustic investigations) to his dissertation, on which he had worked under Wilhelm Wien at the University
of Würzburg.[22] In this study, Hartmann-Kempf investigated the effect
of amplitude on the resonance frequency and damping of tuning forks
and steel springs driven by electric oscillations. Despite the new term,

Hartmann-Kempf's experimental regime remained related to that Helmholtz had set out for acoustics forty years earlier. Around 1900, there was
still a clear distinction between electric and acoustic instruments in physics
textbooks and physical cabinets. The telephone, together with the telegraph, was classified and understood as an electric instrument, not an
acoustic one. Electrically driven tuning forks were generally classified as
acoustic instruments (see, for example, Max Kohl AG 1911). Investigations
in acoustics were still mainly related to music and musical instruments and
dominated by the debate on Helmholtz's theory of combination tones. This
persistent view of acoustics as a science of music became particularly crucial
at the Deutsches Museum in Munich, which opened its doors in 1906. It
had a section on technical acoustics, which comprised only musical instruments (Fuchs 1963, 1).

But a "new acoustics" emerged in this electric age, and Robert Hartmann-
Kempf was at its forefront. Robert was the son of Wilhelm Eugen Hart-
mann, one of the founders of the electrical instrument company E.
Hartmann and Company, which soon changed its name to Hartmann and
Braun. The company's other partner was Wunibald Braun, the brother of
physicist Ferdinand Braun. Robert Hartmann-Kempf used Hartmann and
Braun instruments in his dissertation and joined the company afterward.[23]
Hartmann and Braun specialized in electrical measurement instruments for
science and industry and quickly became a leading company in the field. In
1886, the company introduced an alternating current bridge circuit, including a telephone, to measure the resistance of electrolytes. Wilhelm August
Nippoldt of the Physikalischer Verein (Physical Society) of Frankfurt am
Main had proposed the bridge circuit itself in 1868. Friedrich Kohlrausch,
professor at the University of Würzburg and an expert in electrical precision
measurement, proposed using the telephone instead of an electrodyna-
mometer as a detector of the alternating currents. The bridge was largely
used for testing the proper grounding of lightning rods (see figure 2.3).
Hartmann and Braun designed a small telephone with integrated variable
resistances and a scale that made the apparatus practicable for field use. The
1894 Hartmann and Braun catalog showed two other measurement bridges
with different telephones.[24]

The Hartmann and Braun bridge circuits are only one example of how
the telephone had now become a scientific instrument in its own right
for both electrical and acoustic investigations. In 1901, Hartmann-Kempf

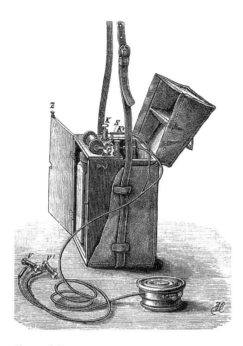

Figure 2.3
Nippoldt telephone bridge for testing proper grounding of lightning rods, by Hartmann and Braun (Hartmann and Braun 1886, 204). The specially designed telephone is the round object at the bottom right. The bridge was not a precision instrument, but it offered a compact and robust design in a carrying box for field measurements by engineers. Hartmann and Braun also offered precision telephone bridges for laboratory use.

conducted experiments for Wilhelm Wien on the acoustic properties of the telephone. The incentive for the experiments came from Frank Sidney Wrinch at the University of Würzburg's Psychology Department. Wrinch had used the telephone in his investigations of the sense of time, producing a continuous tone of a specific duration. Hartmann-Kempf aimed to measure subjective time, and thus had to determine whether the sound of the telephone represented the objective time of the telephone circuit being closed. To this end, he used Hartmann and Braun telephones, designed for electrical measurement, as well as Siemens and Halske telephones, which were the standard telephones of the German postal authority, the Reichspost. The vibration of the telephone membrane was recorded using an optical arrangement whereby a light beam was reflected from a mirror attached to the membrane. The reflected image of the light beam was then recorded photographically on a moving sheet of celluloid film (Kempf-Hartmann 1902).

Wilhelm Wien's cousin Max Wien had already employed an optical mirror attached to the telephone membrane in 1891. Max Wien applied the optical telephone in a bridge circuit for measuring capacitance for alternating currents, which is still known as a Wien bridge circuit (Wien 1891a, 1891b, 1891c). Despite its name, the optical telephone remained an electrical and acoustic technology in Wien's language and practice.[25] While Wien measured the amplitude of the alternating current passing through the telephone optically, he tuned his telephone acoustically. More than any other German scientist of the pre-war era, he embodied the convergence of acoustics, electrical technology, and high-frequency wireless. Wien had studied with Hermann Helmholtz and August Kundt in Berlin. He became professor of physics at the Technische Hochschule in Danzig (today's Gdańsk) in 1904 and moved on to the University of Jena in 1911. Wien employed the telephone in his measurements of the loudness of tone in 1888 and the sensitivity of the human ear for tones of different frequency in 1903 (Wien 1889, 1903). He found the tuning forks used in earlier experiments to be unsuitable for such sensitivity measurements because the amount of energy emitted with the tone could not be determined.

The telephone did not remain the sole electroacoustic technology in the physics laboratory. Around the turn of the century, the singing and speaking arc emerged as a scientific object, connecting electroacoustics directly with wireless technology and the rising popularity of an electromagnetic worldview.

2.5 Electroacoustics in the Electromagnetic Worldview: The Electric Arc as an Experimental System for Oscillation Research

With the rise of electromagnetic theory, electric machinery, and the identification of cathode rays as electrons around 1900, an electromagnetic worldview swiftly supplanted mechanical conceptions, replacing the fundamental laws of mechanics to which all physical phenomena should be reduced with equally fundamental laws of electromagnetism.[26] As Bruce Hunt has put it, "Physicists ceased to feel a need to look for a mechanism behind the electromagnetic laws or to believe that their understanding would be improved by one" (1991, 104). What remained was the view of oscillations as a fundamental principle that transcended and linked all the subdisciplines of the physical sciences. The understanding and mathematical mastery of oscillations was essential not only to electrodynamics but also to the formation of new research fields such as atomic physics and quantum theory. With the rise of the electromagnetic worldview and electrical technology, electric oscillations increasingly replaced mechanical models as a way of understanding oscillation theory and practice.

While the history of the telephone is well remembered, the speaking or singing electric arc has fallen almost into oblivion. Yet around 1900, the electric arc was a widespread technology, with great potential for the future. Its most common form was the carbon arc lamp used for street lighting and projectors. The arc was also used as a radio transmitter for wireless telegraphy, experiments on wireless telephony, and in the electrochemical industry, binding nitrogen from the air for fertilizers and explosives. As an oscillating system, it could broadcast both electromagnetic waves in free space and the human voice without a membrane. The arc could detect electromagnetic waves as well as sound. What could be a more ideal experimental system for acoustics in the electric age?

Carbon arc lamps were supplied with a steady direct current from a battery or a direct-current generator. Hertha Marks Ayrton studied the hissing or humming sound that the arc produced when the current changed rapidly (Ayrton 1902). As one of the few women in electrical research, Hertha Marks was not allowed to take a degree at the University of Cambridge, where she had studied mathematics, but received a bachelor of science degree from the University of London in 1881. In 1884, she started to study electrical engineering and met her later husband William Edward Ayrton, with whom she collaborated in her electrical researches (Mason 2006). William Du Bois Duddell, a student of William Ayrton's at London Central Technical College, investigated the conditions for sound production in the arc circuit, and around 1900, he found that the arc emitted a clearly audible note when an alternating current was superimposed on the direct current (Duddell 1900; see figure 2.4). In fact, Hermann Theodor Simon at the University of Erlangen had made similar observations in 1897. Instead of superimposing an alternating current, he

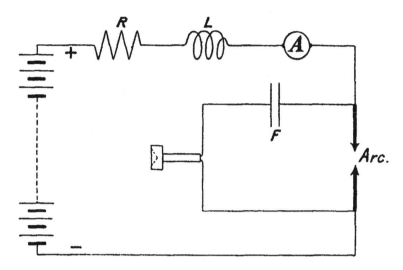

Figure 2.4
Circuit diagram of the speaking arc as a telephone transmitter (Duddell 1900, 242).
The outer circuit supplied the arc with a continuous current. The inner circuit connected the telephone receiver to the arc. Duddell connected the circuit directly, not
by induction, as Simon had done. All sounds near the arc were heard in the telephone receiver of the inner circuit. Duddell found that the transmitted sounds in the
telephone receiver were not very loud and were accompanied by distorting sounds
from the arc.

varied the direct current by introducing a microphone or a telephone into
the current. When Simon spoke or sang into the microphone, the sound
was clearly reproduced by the arc. But there was more: when Simon spoke
or sang into the arc, the pressure changes produced alternations in the
current of the arc circuit. In a second circuit, connected to the arc circuit
by induction, he could then reproduce the sound in a telephone. The arc
could not only reproduce sound; it could also act as a microphone to
detect and transmit sound and could thus be used as a telephone transmitter (Simon 1898).

One of the possible advantages of the arc as a telephone receiver was that
it was reasonably loud, and its sound intensity could be increased by
increasing the length of the arc. No evidence suggests, however, that the
electric arc was ever used as a practical telephone receiver apart from experimental setups and demonstrations in classrooms (see figure 2.5). Duddell
constructed a keyboard arrangement with eight keys, each connecting a

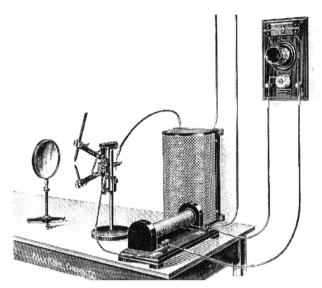

63 395, 50 892, 63 386, 63 388, 63 389.   1 : 8.

Complete Apparatus for Experiments with the Speaking Arc Lamp after Weinhold (W. D., ℓ ≈ d. Fig. 601), consisting of:

Arc Light Hand Regulator and Resistance: see Nos. 50,892, 63,386 and 63,387, p. 1057.

63,390. **Induction Coil,** F i g u r e 3. 4. 0

63,391. **Microphone with Switch and Regulating Resistance, and 1 Fuse with simple current indicator for same** 0. 0. 0

Additional Apparatus for above so as to be able to demonstrate the automatically singing (whistling) arc lamp also, consisting of:

63,392. **Small Induction Coil** (W. D., Fig. 604), with aluminium ring for Thomson's Experiment   0. 16. 0

61,122. **Switch** 0. 3. 6

63,393. **Paper Condenser** in simple wood box, approx. 8 mfd. 2. 0. 0

63,394. **Staged Paper Condenser,** with four steps, for use instead of No. 63,393; this permits of obtaining a simple melody with the arc (electric piano) since the pitch varies with the capacity cut in 6. 0. 0

Figure 2.5

Apparatus for singing and speaking arc experiments from the 1911 Max Kohl catalog (Max Kohl AG 1911, 3:1058). The speaking arc became a popular demonstration device in physics classes: it could be used with a microphone or as an "electric piano," playing a simple melody. In the same section of the catalog, entitled "Telephony and Microphony," Max Kohl advertised sets for photophonic apparatus (wireless light telephony), employing acetylene light and selenium cells.

condenser of a specific capacity to the shunting circuit. This transformed his arc into a musical instrument, which he could play like a piano.

Simon and Duddell used the electric arc to produce and detect audible frequencies only. But the electric arc circuit could also produce much higher frequencies, suitable for wireless telegraphy. Hertz, as well as the Marconi system and other early wireless systems, produced high-frequency electromagnetic waves by high-voltage discharge through a spark gap. This generated relatively short discontinuous pulses of waves over a broad range of frequencies, which were highly damped. To broadcast sound rather than telegraph signals—in other words, to achieve wireless telephony—the wave had to be continuous and undamped. Duddell's musical arc produced such undamped continuous waves. Their frequency, however, had to be of a much greater magnitude.

Neither Duddell nor Simon ended up developing an arc transmitter for wireless telegraphy and telephony. The credit for that accomplishment goes to Danish engineer Valdemar Poulsen.[27] Poulsen and his colleague Peder Oluf Pedersen were able to increase the frequency by burning the arc in hydrogen gas rather than in air, replacing the carbon anode with water-cooled copper, and applying a magnetic field across the arc. The Poulsen arc not only facilitated wireless telephony, but also enabled precise tuning of the transmitting radio frequency. Poulsen set up his first experimental transmitter in 1904, but not until the end of the decade did the arc transmitter manage to break into the fiercely competitive market for wireless systems.[28]

Simon continued to systematically study the electrical and acoustic properties of the electric arc after he was appointed as a professor at the Institute of Physics, University of Göttingen, in 1901.[29] He succeeded Theodor des Coudres as the chair of Applied Electricity and became head of the division with the same name. Under the directorship of des Coudres, teaching and research in the division had focused on heavy-current electrical engineering; Simon shifted that agenda toward wireless telegraphy, electromagnetic oscillations, and gas discharge. He labeled the scientific field that ultimately brought all these activities together "applied ether physics" (*angewandte Aetherphysik*).[30] In 1905, Simon's division became a separate Institute for Applied Electricity within the larger Institute of Physics, and Simon was able to inaugurate his own building next to the new physics building (Simon 1906). A Division of Technical Physics had been founded in 1897, which focused mainly on applied mechanics. When Ludwig Prandtl was appointed chair of Technical Physics in 1904, this division was combined with the Division of Applied Mathematics under the Institute for Applied Mathematics and Mechanics.

The institutes for Applied Electricity and Applied Mathematics and Mechanics were part of Felix Klein's efforts to establish the applied sciences as part of the sciences at the university. Klein was one of Germany's most influential mathematicians. Appointed in Göttingen in 1886, he became an effective science organizer who helped to make the University of Göttingen an international center for mathematics and the physical sciences. Klein wanted to bring together universities and Technische Hochschulen and to connect pure science, applied science, and technology in teaching and research. He had traveled to the United States in 1893 and 1896 and had found American universities a model of close collaboration among science, technology, and industry. Klein was one of few university professors who supported the claims of the Technische Hochschulen to grant doctoral degrees. As a result of Klein's activities, the Göttinger Vereinigung zur För- derung der angewandten Physik (Göttingen Association for the Advance- ment of Applied Physics) was founded in 1898 by leading members of German industry, including the electrical company Siemens and Halske and the Friedrich Krupp steel manufacturer, as well as by representatives of the University of Göttingen. The role of the association was to procure funding for the Göttingen laboratories and to maintain close contact between applied physics and industry.[31]

At the Institute for Applied Electricity, Simon transformed the electric arc into an experimental system, making it the core of his research agenda.[32] Many of his assistants and doctoral students carried out experimental and theoretical investigations of different aspects and characteristics of the arc. Among them were Max Reich, Heinrich Barkhausen, Karl Willy Wagner, Wilhelm Rihl, Hugo Lichte, and Gertrud Lange.[33] Rihl and Lichte worked specifically on the sound intensity of the electric arc (Rihl 1911; Lichte 1913).[34] Whether the sound of the arc was produced by the heat variations of the oscillating arc or by an evaporation of the carbon electrodes still had to be determined; for technical applications, it was more important to study how the sound intensity could be increased most effectively.

Heinrich Barkhausen's thesis, *Das Problem der Schwingungserzeugung mit besonderer Berücksichtigung schneller elektrischer Schwingungen* (1907; The problem of generating oscillations with special consideration of rapid electric oscillations), was arguably the most influential and lasting contribution to come out of Simon's institute. In Barkhausen's comprehensive study of the generation of oscillations in electrical and mechanical systems, he investigated the analogies between mechanical and acoustic vibrations on the one hand and electric vibrations on the other, taking electric oscillations as a point of departure. Barkhausen's approach was systematic and symmetrical. The generation of undamped continuous high-frequency electric oscillation was crucial for wireless telegraphy, and even more so for wireless telephony, which was still at an experimental stage. The amplification of electric oscillations, in turn, was vital to long-distance telephony. Barkhausen did not place the electric arc at the center of his study, treating it instead as a special case for the generation of high-frequency electric oscillations (Barkhausen 1907). Barkhausen's analog treatment of electric, acoustic, and mechanical oscillations was the origin of his research program, Schwingungsforschung (oscillation and vibration research), developed after he was appointed professor of electrical engineering at the Technische Hochschule in Dresden in 1911. More importantly, Karl Willy Wagner, who had studied the electric arc as a generator for alternating current in Göttingen, was to adopt the program of oscillation and vibration research as the defining principle of the Heinrich Hertz Institute for Oscillation Research, which he would found in Berlin in 1927. The symmetrical and analog treatment of electric and acoustic oscillation thus became the core concept of electroacoustics.

The consolidated efforts of Simon and his students to investigate the electric arc as an experimental system allowed Simon to present a synthesis of its current state and future prospects as a technology, as well as to locate it within contemporary developments in general physics, for which the University of Göttingen's Institute of Physics was a leading center. In an experimental lecture in Berlin in 1911, Simon gave an overview of the different properties and potential capabilities of the arc, including its heat- and light-producing powers, its capacity to produce and consume sound, and finally its capacity to produce and detect rapid electric oscillations in free space (Simon 1911). Simon did not stop at sketching the technical prospects of the arc as a simple yet omnipotent wireless electroacoustic communication system. His goals as a physicist were never merely technological—he intended to ground technology in physics. He placed the electric arc as applied ether physics firmly within the electromagnetic worldview. Even if that worldview was already declining in 1911, Simon still saw it as the dominant one in the scientific community. The building blocks of Simon's electromagnetic view were electrons, atoms, and the all-penetrating world ether (Simon 1911, 32–35). The electric arc as a stream of ions between the electrodes fitted neatly into this picture. Through their research on the electric arc as an oscillating system, Simon and his students placed acoustics, perhaps for the first time, in an electrical frame rather than in a mechanical one.

Around 1906 the Reichsmarineamt (Imperial Naval Office) approached Simon for a partnership. The physicist H. Ament of the German Navy's Torpedo-Versuchs-Kommando (Torpedo Testing Command; TVK) came to work in Simon's laboratory on an experimental investigation into the generation of undamped electric oscillations.[35] This was only the beginning of Simon's collaboration with the German Navy. In 1908, he set up the Radioelektrische Versuchsanstalt für Heer und Marine (Army and Navy Radioelectric Testing Laboratory), which was affiliated with the Institute for Applied Electricity but subordinated to and financed by the TVK. The laboratory consisted of three radio masts (each 80 meters high), a building with three rooms, and equipment to measure the propagation of the electric waves. Max Reich, a reserve lieutenant, was appointed head of the testing laboratory in 1909.[36] The Radioelectric Testing Laboratory carried out training courses in wireless telegraphy for officers of the German Army and Navy as well as for schoolteachers. It was no coincidence that both Simon and many of his former students—including Max Reich, Heinrich Barkhausen, and Hugo Lichte—came to work for the TVK during World War I.[37]

In this chapter I have discussed the main forces and actors that fueled the transformation of fin de siècle acoustics research. Luigi Russolo's futurist manifesto forecasted the symphony of the industrial war and challenged bourgeois society as the old world order. Between the turn of the century and the outbreak of World War I, few physicists engaged in traditional acoustics research as laid out by Helmholtz and Rayleigh. A new type of acoustics driven by the rise of electric communication technology was already under way and would unfold further under the constrained conditions of the war. Simon's institute in Göttingen was the ideal place to lay the foundations of the new acoustics research agenda, as Barkhausen did in his dissertation. The electric arc as an experimental system combined the frontiers of physics with applied and industrial research, just as Felix Klein had envisioned. With the war, radio valve transmitters quickly dominated wireless technology. The valves used in radio transmitters also amplified signals from microphones and telephones. The speaking or singing arc disappeared into the cabinet of curiosities of electroacoustic technologies.

3 Science Goes to War: Warfare and the Industrialization of Acoustics

3.1 Acoustics in the Chemists' War

World War I broke out on 28 July 1914 and lasted until 11 November 1918. It brought an end to the bourgeois and aristocratic world order of the Wilhelmine and Edwardian era. Though its main battles were carried out in Europe, World War I was a global war to the extent that it affected most of the world's population, including many people in the colonies of the European imperial powers. The war dramatically changed the political, economic, and cultural landscape on the European continent, including the relationship between science, industry, and the military. The outcome of World War I cast the bourgeois scientific establishment in Germany into a crisis from which it did not recover throughout the Weimar Republic.[1]

As I argue in the previous chapter, the conception of acoustics as bourgeois science par excellence, defined by classical music as high culture, had been challenged before 1914. But the industrialization of acoustics and its emergence as an archetype of industrial physics or technical physics took place only during World War I. Historians of acoustics have repeatedly asserted that acoustics was perceived as an old-fashioned and outdated research field in the decades between 1890 and the outbreak of the war, and that the war was the main driving force for the revival of acoustics research in the early twentieth century.[2] The notion of Technische Akustik (technical acoustics) had already been used before the war, but acoustics was still a science of music, and its technology a musical technology. Erich Waetzmann's *Die Resonanztheorie des Hörens* (The resonance theory of listening) of 1912 was thoroughly rooted in Helmholtz's framework. It is revealing, for example, to look at how acoustics was presented in the large European museums of science and technology in the period. At the opening of the Deutsches Museum in Munich in 1906, the section on technical acoustics included musical instruments only—telephony and telegraphy were in a different section. Myles Jackson and Sonja Petersen have argued that musical instrument manufacture was standardized and industrialized during the nineteenth century (Jackson 2006; Petersen 2011). But the image of musical instrument making and music performance continued to be one of craftsmanship and virtuosity, as it remains even today. The kinds of artifacts and aesthetics produced by the battlefields of World War I were substantially different. In 1921, Ernest Lancaster Jones of the Science Museum in South Kensington collected and exhibited hydrophones for U-boat detection, an aircraft locator, the Bull sound-ranging apparatus for artillery ranging, and a Tucker hot-wire microphone.[3] These exhibits at the Science Museum contrasted the prewar exhibits of technical acoustics as a technology of music with the acoustics of industrial warfare.

World War I is often called the Chemists' War, emphasizing the importance of the chemical military-industrial-scientific complex in producing substitutes for scarce raw materials and, of course, chemical warfare. Yet, the epithet underplays the role of physicists and mathematicians in the mobilization and self-mobilization of scientists for war-related work, both on the battlefield and in the laboratories behind the front (Schirrmacher 2009, 155). World War I was the first large-scale industrial war in which weapon systems like the submarine, airplane, and tank were used, and for which large numbers of scientists from all disciplines were deployed in warfare research and development (Trischler 1996; Hartcup 1988, 68–80, 129–135, 163–164; Hackmann 1984; Schirrmacher 2009, 2014). At the beginning of the war, the military establishments and the public in France, Britain, and Germany believed that the conflict would last only a few months, until the end of 1914. Military leaders did not call for research and development efforts by civilian physicists, who were enlisted as soldiers and sent to the front, at most to take on engineering tasks, such as operating wireless stations or X-ray apparatus in field hospitals. The situation changed in 1915, with gridlocked trench warfare and the rapid development and growing importance of aircraft and U-boats. The initiative for research and development projects came mostly from scientists approaching military authorities rather than the military approaching scientists. At the same time, physics departments were emptied of students and staff, and experimental equipment was difficult to procure (Wolff 2006).

Before the war, most physics research in Germany was institutionally located at the universities, in contrast to chemical research, much of which was carried out in the chemical industry in close collaboration with university chemists (Reinhardt 1997). Compared to physics, chemistry also had a much stronger presence as a discipline at the Technische Hochschulen.[4] In 1911, a new player entered the institutional landscape of science in Wilhelmine Germany: the Kaiser-Wilhelm-Gesellschaft zur Förderung der Wissenschaften (Kaiser Wilhelm Society for the Advancement of Science), which became instrumental in the organization of the German scientific war effort. In the Kaiser Wilhelm Society, it was again chemistry that took the lead. The first two institutes founded by the society were the Kaiser Wilhelm Institutes (KWIs) for Chemistry and for Physical Chemistry and Electrochemistry. Fritz Haber, architect of the chemistry war effort and gas war, became director of the KWI for Physical Chemistry and Electrochemistry (Szöllösi-Janze 1998). With Albrecht Schmidt of the chemical dye company Hoechst and senior governmental official Friedrich Schmidt-Ott, Haber initiated the Kaiser Wilhelm Stiftung für Kriegstechnische Wissenschaft (KWKW; Kaiser Wilhelm Foundation for Military-Technical Science), which was founded in October 1916 and began research activities in spring 1917 (Rasch 2006).

Compared to the German chemists and chemical industry, the war effort of the German physicists was rather scattered and without central organization. Though there was a physics committee at the KWKW, chaired by Walther Nernst, its activities seem to have been limited, and its influence on postwar developments in technical physics negligible.[5] The Physikalisch-Technische Reichsanstalt was mainly occupied with testing activities for the military and did not conduct war-related research and development (Cahan 1989, 225–226; Scheel 1918, 1919). Few departments of physics at German universities were oriented on technical physics. Among these were the Institutes for Applied Electricity and Applied Mechanics, directed by Hermann Theodor Simon and Ludwig Prandtl, respectively, at Göttingen University, and the Department of Physics at the University of Jena, where Max Wien had been appointed professor in 1911.[6]

Scientists and military officials established institutions similar to the KWKW to coordinate research and development by civilian researchers for the war effort in Britain, France, and the United States. The most important British institutions were the Board of Invention and Research, set up by the British Admiralty in July 1915, and the British government's Department of Scientific and Industrial Research (DSIR), launched in December 1916. The Board of Invention and Research brought together some of Britain's leading physicists, including J. J. Thompson, Ernest Rutherford, and William Henry Bragg. Because of tensions between the civilian scientists and Royal Navy officers, the board was dissolved as early as 1917. In France, the Ministry of Education founded the Direction des inventions in 1915, which was later transferred to the Ministry of Munitions (Rasmussen 2010, 311). The United States came in relatively late, having declared war only in February 1917. The National Research Council (NRC) was established in 1916 and headed the coordination of war-related research under the leadership of George E. Hale and Robert A. Millikan (Kevles 1978, 102–138).

The war was not fought only in the trenches. Aircraft took the conflict into the air and far behind enemy lines; submarines took it below the surface of the sea. Heavy artillery enabled precision shelling far beyond the front line. Tunnels dug under the enemy's trenches moved fighting underground. Scientists and engineers developed and improved methods and technology for electric communication such as telegraphy, telephony, wireless, and underwater sound signaling. Communication was essential for this global industrial war, whether it was communicating with colonial administrators thousands of kilometers away, between ships and aircraft, or from the command center down into the trenches and back. All armies used field telephones in the trenches.[7] Telephone lines could be cut, however, and indeed the telegraph lines to the German colonies were destroyed, giving rise to demands for less vulnerable wireless communication.

Scientists' horizons with respect to the scope and measurement of acoustics were expanded mainly through the sound location of enemy artillery, aircraft, U-boats, and tunnels (Hartcup 1988; Encke 2006; C. Hoffmann 1994; MacLeod 2000; Hackmann 1984). The objective of all sound location—whether in the air, underground, or in the ocean—was to gather information about the enemy. The sound of artillery shells, aircraft and ships could reveal information about their precise location and activities. The same knowledge that was used to detect the enemy was also used to silence an army's own aircraft and ships, thus concealing their activities. As it turned out, the wartime development of wireless telegraphy and telephony would become equally important to electroacoustics in the interwar period. While acoustics in the battlefield became increasingly electrified, wireless telegraphy became increasingly acoustical. For all three—telephony, wireless telegraphy, and sound detection—the development of amplifier tubes and knowledge about electric circuits were crucial. In this sense, World War I acted as a large industrial laboratory for acousticians, who now turned to new sounds, new techniques, and a new understanding of their discipline.

3.2   Signal and Noise: Sound Detection on the Battlefield

In the war years, when the artillery was thundering outside, it was all the quieter on the streets and in the scientific institutes in Berlin. Acoustic observations could be made almost as in a soundproof room, and even whispering sounds could be analyzed down to their last component. (Stumpf 1926, v)[8]

In 1905, Erich Moritz von Hornbostel, Carl Stumpf's disciple and the director of the Berlin Phonogram Archive, had advocated for the value of "atomizing" the music of the Herero and other Khoikhoi tribes of German South West Africa into its component parts, with scientific precision, by using the tonometer and metronome (Hornbostel 1905, 85). Ten years later, Hornbostel and his fellow acousticians analyzed the sounds of guns and machines rather than pieces of music. The scientific foundations for the sound ranging of artillery were known before World War I, but only with the beginning of trench warfare did the military leadership consider putting them into practice. Most military leaders had not anticipated the implications of the heavy artillery and machine guns that brought moving armies to a halt. The character of naval warfare and the role of aircraft, too, changed rapidly and in unpredicted ways during the course of the war. German U-boats became a threat to the British navy and merchant ships, while aircraft took on an increasingly active role in combat. Both U-boats and aircraft could be located and identified by the sounds they created.

In the first years of the war, human observers aided by stopwatches, trumpets, microphones, and hydrophones carried out all sound detection in the air, under water, and in the tunnels. Sound detection consisted of analytical and interpretive listening to the roar of the battlefield. The task of the observer in a sound-ranging section was to distinguish among all the disturbing noises of battle the sound of the one gun that was to be observed. The various (battle)fields of sound detection—artillery, aircraft, submarine, and tunnels—required different strategies, but the principal goal was the same: to locate the enemy by its sounds and to gather information about enemy machines and activities by analyzing those sounds.

Sound detection became a technology within large and complex technological systems: military surveying, submarine warfare, and aerial warfare.[9] The acoustic knowledge and practices of sound detection transcended the boundaries of these otherwise largely separate systems and established links between them. Sound location triggered measures to detect and to avoid detection. Well aware of the activities and methods of the other side, the different actors of the belligerent powers were developing not only acoustic methods to locate the enemy and analyze its activities, but also counter-methods to acoustically hide their own activities. This could mean silencing machinery such as aircraft and U-boats, or hiding their sounds in the clamour of the battlefield.

Scientists on both sides worked feverishly on automating sound detection in order to eliminate the need for human observers and all the problems they posed for precision sound measurement.[10] Sound detection by human observers was, by definition, a subjective method that depended on the hearing, sound perception, and reaction time of the individual. The lack of objectivity in human sound observers had already been criticized by Stumpf in the field of comparative musicology, where Stumpf and his disciples claimed to have established objectivity by means of phonography. For the location of the very specific sound signals on the battlefield, however, the phonograph was not suitable. The tools of objective sound recording changed from the phonograph, tonometer, and metronome to microphones, string galvanometers, oscillographs, and photographic paper—but the idea of objective sound measurement relying on sound recording and its quantitative analysis remained the same.

3.2.1 Artillery Ranging

The rapid movement of the German, French, and British armies stopped at the western front in the autumn of 1914. The static front went into trench warfare, the armies digging themselves deeper and deeper into positions attacked by the precision shelling of heavy artillery. Precision shelling required more exact mapping of the front line to supply data for targets (MacLeod 2000, 26–34). At the same time, heavy artillery itself was a target for precision shelling. Sound ranging became part of a complex system of surveying that included mapping, aerial photography, flash spotting, meteorological measurements, and mathematical analysis, all brought together by a "compilation section." As a countermeasure, the armies increasingly used techniques to hide their artillery guns from sound as well as visual detection. Acousticians analyzed not the machine-gun "ta-ta-ta" of Filippo Marinetti's poem on the Battle of Adrianople, but the low-frequency "boom" of heavy artillery, making sound ranging an activity of high-precision measurement.

Sound ranging of artillery did not have a unique inventor but was proposed more or less simultaneously in 1914 by scientists, artillery officers, and surveyors of the French, German, British, and Austrian armies. On all sides, the initial proposals included subjective as well as objective methods of sound ranging to detect enemy artillery. Many factors affected the accuracy of sound ranging, with wind speed, wind direction, and temperature all needing to be measured and included in the calculation. During low visibility, sound ranging was the only method to determine location, while certain wind directions rendered it useless. Fast-moving armies hindered sound ranging, since the sections needed time to take their positions, set up the equipment, and adapt to the new territory.

The sound-ranging activities of the German Army were coordinated by the Artillerie-Prüfungs-Kommission (APK; Artillery Testing Commission). The APK was responsible for all development and testing of artillery material. It operated a large firing range at Kummersdorf outside Berlin, and other testing and training facilities, such as the artillery-surveying school in Wahn. Under the military command of Captain Fritz von Jagwitz, the Breslau physics *Privatdozent* (adjunct professor) Rudolf Ladenburg was the scientific head of work on sound ranging at the APK. According to R. Harbeck, Jagwitz worked with twenty-four scientists and a total staff of three hundred, of whom "all were academically trained" (Harbeck 1943a, 9). At Ladenburg's suggestion, Max Born—professor extraordinarius in Berlin at the time—was hired as a theoretical physicist to verify the experimental data of sound ranging and calculate corrections for wind speed and direction.[11] Among the physicists developing methods and instruments for sound ranging and the analysis of data were the University of Göttingen Privatdozent Erwin Mandelung, who had been a doctoral student of Simon's; Karl Fredenhagen, professor extraordinarius of physics and physical chemistry at the University of Leipzig, who had been Simon's assistant in Göttingen; and Ernst von Angerer, curator at the Physics Department of the Technische Hochschule Munich.[12]

Early sound-ranging sections consisted of several human observers equipped with stopwatches to record the time when the firing of a gun was heard at specific locations. Physicist and chemist Leo Löwenstein proposed this method in 1913. Both Fredenhagen and Löwenstein had completed their doctorates at the University of Göttingen with Walther Nernst, who chaired the physics committee at the Kaiser Wilhelm Foundation for Military-Technical Science. The German Army introduced Löwenstein's method at the front in May 1915 (Löwenstein 1928; Menges 1987). By calculating the time differences between the locations and using triangulation, sound-ranging sections could determine the location of the gun. The critical factor for the accuracy of these subjective sound-ranging systems was the human observer. The problem of synchronizing stopwatches could be overcome by electric timekeeping, but a more serious obstacle was the observers' inherent long and uneven reaction time.[13] Accordingly, the scientific team at the APK testing commission included experimental psychologists, among them three former students and collaborators of Carl Stumpf: Hans Rupp, Erich Moritz von Hornbostel, and Max Wertheimer.

Rupp, who had been Stumpf's assistant since 1907, started his experiments at the APK in Kummersdorf in January 1915 and soon extended them to the front line. Rupp studied those subjective reactions of the human observers that were relevant to sound ranging. He measured variations in the observers' reactions and tested potential candidates for work in sound-ranging sections (C. Hoffmann 1994, 265; Rupp 1921b). In October 1916, as an Austrian citizen, Rupp was drafted into the Austro-Hungarian Army, where he continued to select and train personnel for sound-ranging sections. Hornbostel and Wertheimer developed a different subjective method, which was based not on determining time differences but on the binaural hearing sense of observers—binaural hearing made it possible to locate the direction from which the sound of the gun had come. The two men carried out some of their experiments at Stumpf's Psychological

Institute in Berlin, where Wolfgang Köhler had worked on binaural hearing prior to World War I.[14]

The precision of the binaural hearing method depended not on the reaction time of the observers but on how precisely they could determine the direction of the gun they had heard. To ascertain the direction of a gun being fired with adequate precision, the observers used stethoscopic devices that extended the distance between the left and right ear. A trumpet or horn at each end of the stethoscope collected sound volume, which was then directed to the relevant ear through tubes. Köhler had used a similar device in his experiments before the war. The German Army introduced the binaural sound locator, designed and patented by Hornbostel and Wertheimer (figures 3.1 and 3.2), and used it in tandem with the Löwenstein stopwatch method.[15] In the French Army, the binaural sound locator was known as the "orthophone." Although the French and British armies seem to have quickly abandoned binaural sound locators for artillery ranging (Winterbotham 1918, 33–34), similar sound locators were developed for aircraft detection. Binaural hearing methods were also used in the underwater sound location of U-boats. I come back to both points later.

The obvious disadvantages of the subjective method using the Hornbostel and Wertheimer sound locator were, first, its lack of precision because of the human factor, and second, the low range of possible observations and the high risk of the observers being seen and shot at.[16] Measurements could not be cross-checked later, as was possible with the recordings of automated apparatus. The French and British armies, who collaborated in their sound ranging, also evaluated and tested several different objective and subjective methods. Both armies chose to develop an objective approach (MacLeod 2000, 34–36). Central to French sound-ranging efforts were the astronomer Charles Nordmann of the Paris Observatory and the Irishman Lucien Bull of the Institut Marey. Bull had used string galvanometers for electrocardiography before the war (Van der Kloot 2005, 275), and now he developed a system whereby an Einthoven string galvanometer registered the sound signals of up to seven microphones. The displacement of the strings by the microphone current was recorded on a moving strip of photographic paper. A tuning fork driven by an electromagnet provided a precise time reference that was recorded on the same paper (see figure 3.3). The whole system was

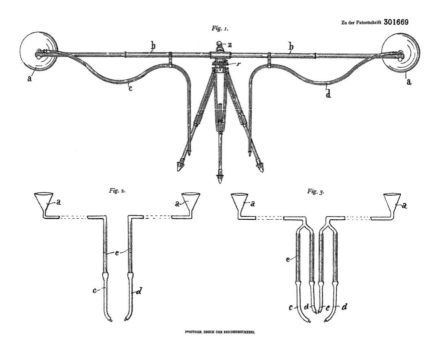

Figure 3.1
Drawing of Erich Moritz von Hornbostel and Max Wertheimer's sound locator for the German Army, taken from their patent specifications filed on 6 July 1915. The method was based on increasing the precision of the sense of direction among sound-ranging observers. The distance between the human ears was increased by the trumpet arrangement. The trumpets *a* collected the sound, which was directed to the observer's ears by the tubes *c* and *d*. The trumpets' horizontal base *b* was then turned around the axis *z* until the observer perceived the sound as originating not from left or right but directly from the front. The angle needed for triangulation could then be read at the aiming circle *r*. The locator at bottom left shows the arrangement for one observer, and at bottom right, the arrangement for two observers simultaneously to increase the reliability of observation.

set in operation by forward observers ahead of the microphones, who gave the starting signal by telephone. Once the photographic paper was developed, the time difference between the signals of the various microphones was determined. From this, depending on the location of the microphones and after wind speed and temperature corrections, the location of the gun could be identified.

A first sound-ranging section with Bull's automatic system was sent to the battlefield in January 1915 (Winterbotham 1918, 32). A British Army

Figure 3.2
The German Army's Hornbostel-Wertheimer sound locator in action. The two additional trumpets on the left below and above the base were for height location. The sound locators had to differentiate between the sound originating from the muzzle of the gun and the sound of a supersonic shell originating from its trajectory higher up (Bochow 1933, plate opposite p. 32).

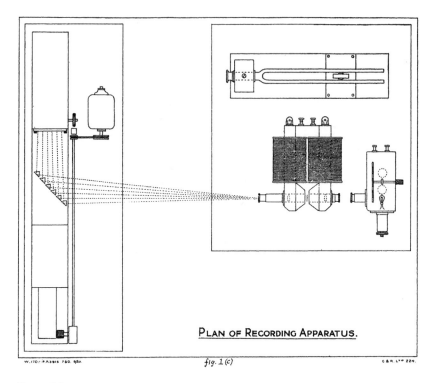

PLAN OF RECORDING APPARATUS.

fig. 1 (c)

Figure 3.3
Plan of the Bull sound-ranging apparatus used by the British Army's sound-ranging sections (*Handbook of the Sound Ranging Instrument* 1921, fig. 1c). On the right: lamp arrangement and Einthoven string galvanometer. Each of six microphones was connected to one of the galvanometer's six strings. When a current passed through the strings, the light beams (shown as dotted lines) were deflected. The electrically driven tuning fork at the top right controlled the time wheel, which served as a chronometer. On the left: the actual recording apparatus. The light beams from the galvanometer were reflected against the silvered glass prisms and directed onto the photographic paper. The paper was moved by an electric motor. A light beam from the time wheel added marks for the time reference.

committee investigated the prospects of sound ranging on the French front lines in spring 1915. They judged the Bull system, still at an experimental stage, to be the most promising, and decided to adopt it for their own sound-ranging sections. In the summer of 1915, the British Army seconded the Australian physicist William Lawrence Bragg, aged just twenty-five, to head a group of six officers and other men, among them a mathematician, an electrician, and an instrument maker (Jack 1920, 106). Despite his youth, W. L. Bragg was a renowned scientist from the Cavendish Laboratory. With his father, William Henry Bragg, professor of physics at Leeds and later at University College London, William Lawrence was awarded the Nobel Prize in Physics on 12 November 1915 for research on the nature of X-rays and their application in determining chemical structures (Van der Kloot 2005). This made W. L. Bragg the youngest Nobel laureate ever.

The initial results of sound ranging with the Bull system were not satisfactory. The artillery guns produced sound waves of very low frequency, whereas the conventional telephonic microphones used by the French Army were optimized, if at all, to pick up and transmit the human voice.[17] The sound-ranging system would record soldiers talking and rifle shots, but seldom artillery guns (Winterbotham 1918, 34). The problem was exacerbated when the German Army introduced high-velocity guns: moving at supersonic speed, the grenades produced a shock wave that traveled ahead of the sound wave originating from the firing of the gun; this shock wave destroyed the reading from the firing gun. In the search for the ideal sound-ranging microphone, which would pick up only the gun to be detected, the breakthrough came from William S. Tucker, an Imperial College physicist who had obtained his doctorate in 1915. In summer 1916, Tucker experimented with a hot-wire microphone that only recorded the low frequencies of a gun being fired and remained undisturbed by any sound of higher frequencies (figure 3.4). The microphone consisted of a Helmholtz resonator, basically a box resonating in the low frequency to be observed, and a platinum wire mounted over the hole of the resonator. With resonance, the passing air cooled the platinum hot wire and lowered its resistance (Tucker and Paris 1921).

After the introduction of the Tucker hot-wire microphone, sound ranging with the Bull system became very successful at locating enemy guns. Certain wind conditions, however, could still disable sound ranging completely. Like the German Army, the British Army operated several experimental sites and sections for sound ranging throughout the war, as well as schools to train the various sound-ranging sections. The apparatus for the first twenty sections was manufactured at the Institut Marey under Bull's supervision. To speed up production, the Cambridge Scientific Instrument Company was put in charge of producing the instrument sets in 1917. An

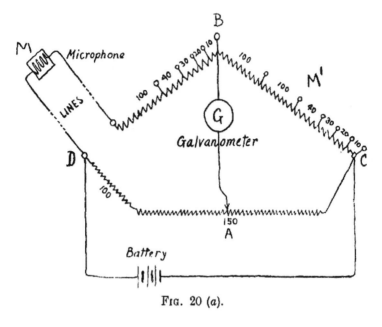

Fɪɢ. 20 (a).

Figure 3.4
Wiring of the Tucker microphone with a Wheatstone bridge (*Handbook of the Sound Ranging Instrument* 1921, 23). At the beginning of the measurement, no current passed through the Einthoven string galvanometer G. When a gun was fired, the air stream cooled the wire of the microphone M. The resistance of the wire changed, and a small current passed through G.

automatic developer for the photographic film was introduced later that year, and the Bull apparatus was constantly improved during the war. The first sound-ranging section, under W. L. Bragg, remained the experimental section. Sound-ranging officers did not need to be scientists, but one member of every section was required to have some background in physics, one some knowledge of electrical engineering, and two had to be trained in basic surveying methods. The commander of the sound-ranging section needed to be a good organizer rather than a scientist or engineer. Listening skills, in contrast, did not seem to be a particular priority (Winterbotham 1918, 41). The British Army maintained a workshop under a specially selected officer, Lieutenant Hereward Lester Cooke, a physics professor at Princeton University and former student of Ernest Rutherford at McGill, with a permanent staff of sixteen craftsmen (Jack 1920, 108). The workshop tested and adjusted new apparatus and repaired and modernized older models as the method became increasingly sophisticated.

Like their British and French counterparts, German scientists had proposed objective methods for sound ranging from the beginning of the war. These proposals included graphical methods to register sound picked up by microphones either on a smoked paper recorder or with oscillographs or string galvanometers on photographic paper. The main drawbacks of objective sound-ranging methods were obvious to many German military officials: compared to the subjective system with stopwatches and the Hornbostel and Wertheimer sound locator, the apparatus proposed was expensive, heavy, and complicated to operate. Even if it worked under laboratory conditions, how could such a delicate measurement system be feasible on the battlefield in trench warfare? Trained human observers could easily distinguish the sounds of different guns by their characteristic timbre, whereas the microphones registered rifle shots and conversation, and often remained deaf to the low frequencies of large-caliber guns. Löwenstein had presented an electric recording and registration system as an alternative to human observers in his original proposal of 1913, and while on military leave in December 1914, Jonathan Zenneck, professor of physics at the Technische Hochschule Munich, proposed an objective method employing a string galvanometer. Karl Wolff then designed the registration apparatus under the supervision of Ernst von Angerer. Parts of the apparatus were adopted from an existing design by the Munich scientific instrument company Edelmann, which later manufactured the instrument.[18] The German Army, represented by the deputy chairman of the APK, General Anton von Kersting, evaluated the apparatus and method and found it unsuited for use on the battlefield (Harbeck 1943a, 15). In January 1915, Karl Fredenhagen proposed a different objective method of sound ranging, employing an oscillograph (Harbeck 1943a, 36–38). Despite inconclusive and negative results, the German Army facilitated Fredenhagen's work, which employed a Siemens and Halske oscillograph and Siemens microphones.[19]

In December 1917, the German Army gained firsthand information about the objective sound-ranging system of the British Army and its allies. German Army intelligence had captured secret manuals from British soldiers, dated March 1917, which were then studied in great detail.[20] German officers recognized that the British system owed its success mainly to a microphone that reacted only to the low frequencies of the artillery guns. The German Army had not been able to capture a microphone, and the manuals gave no details on the design of the Tucker microphone. As German officials started to acknowledge that the British and French sound ranging was more successful than theirs, the German Army stepped up its work on developing its own objective system. The Allgemeines Kriegsdepartement (General War Department) agreed to continue experiments with a smoked paper recorder, a string galvanometer as used in Angerer's apparatus, and the Siemens oscillograph as used by Fredenhagen.[21] Fredenhagen's apparatus was still considered too laborious, complex, and expensive to justify its advantage over the existing sound-ranging methods, and the sound-ranging section of the artillery-surveying school in Wahn saw Fredenhagen's method as impracticable.[22]

The German Army's objective sound-ranging system came to the battlefield comparatively late. By the end of the war, it did not seem to have succeeded in replicating the success of the Franco-British system, which had also been adopted by the American Expeditionary Force (Kevles 1978, 126–129). In April 1918, the German Army finally managed to lay its hands on a Tucker microphone to study from a British sound-ranging section.[23] Richard Berger's lectures on sound ranging, published by the artillery-surveying school in July 1918, discussed all the various subjective and objective methods and types of registration apparatus, which indicates that all were still in use (Berger 1918). Harbeck concluded that only five oscillographs had been delivered to sound-ranging sections in October 1917. Additional oscillographs were ordered in August and October 1918 but not delivered before the war ended (Harbeck 1943a, 44, 46).

3.2.2 Locating Aircraft by Sound

Methods of aircraft sound location were conceived largely from sound ranging for artillery. As sound ranging was part of a larger system of artillery surveying, sound location of aircraft was part of larger antiaircraft systems, including antiaircraft artillery, fighter planes armed with machine guns, and barrage balloons. During World War I, the nature and importance of aircraft changed dramatically from an instrument of surveillance to an active weapon. Bombing raids increased in frequency and started to target civilians. In the first years of the war, the German Army carried out rather ineffective air attacks from slow and vulnerable airships. In 1917, it started to bomb London with long-range bombers, the "G-planes," commonly known as Gothas, built by Gothaer Waggonfabrik (Van der Kloot 2011, 405). As the threat of bomber and fighter planes grew, antiaircraft measures were developed in response.

Scientists' efforts to find methods for aircraft sound location went in two different directions: short-range location and long-range detection. Short-range sound location was intended to detect bombers in the sky during night raids. During daytime and with good visibility, observers could identify and target enemy aircraft by eyesight, but at night planes could only be spotted with the help of searchlights. Once the sound locators had determined the direction with sufficient precision, searchlights would be trained on the planes, which could then be aimed at directly by anti-aircraft artillery and fighter planes. Long-range sound detection, in contrast, was conceived as an early warning system to detect an aircraft attack, day or night, before the planes could be observed visually.

German efforts in aircraft sound location were, again, directed by the Artillery Testing Commission. Compared with British work toward the same goal, the amount of source material on the German side is rather meager. It seems, however, that Erich Waetzmann was central to German sound location. In a report of 3 May 1918, he presented observations and recommendations on aircraft location by sound.[24] Waetzmann's practices of listening for aircraft locations resembled those he had set out in *Resonanztheorie des Hörens* of 1912. The style of the text recalls an account of musical listening rather than a scientific military report.[25]

Waetzmann emphasized that sound location of aircraft needed experienced listeners who could differentiate between the various, barely audible noises. The listener had to be able to "hear the grass grow."[26] Listening to aircraft at night, when it could not be spotted easily by sight, was especially important. Hearing at night was significantly different from hearing in daytime, and listeners had to practice attentive listening in the open night air.[27] Because they had to be protected from ambient noise, Waetzmann recommended as an ideal listening post a pit 1.5 meters deep, which would shield the listener from the immediate surroundings and serve as an amplifying funnel. Valleys, city courtyards, and archways could also act as amplifying listening posts and shields against wind and ambient noise.[28] Familiarity with the characteristic sounds of different aircraft was important, not only to distinguish them from other sounds but also to differentiate between German and enemy aircraft. Most characteristic were motor and propeller sounds, which Waetzmann described as rather harmonic. Specifically, he found certain types of enemy aircraft higher, brighter, and more musical in tone. Waetzmann described the sounds using musical language and tried to identify their musical notes;[29] the listeners then had to memorize these specific sound impressions.[30]

Like the British, French, Italian, and U.S. armies, the German Army developed trumpet apparatus for detecting aircraft (Bloor 2000). For normal aircraft noise, with engines at full rotation speed, the low tones were especially important, which required the funnels to be larger. The APK tested single trumpets 2 to 3 meters long as well as 1.8-meter double trumpets, manufactured by Siemens and Halske and the Auer-Gesellschaft, a company whose main business was gaslight.[31] Waetzmann put particular emphasis on the binaural double trumpet, which extended the base for direction finding and allowed the operator to determine the aircraft's altitude by positioning the funnels vertically (figure 3.5).[32] The double trumpets were similar to the binaural direction finder for artillery ranging (see figure 3.2), but much larger.

Figure 3.5
The Waetzmann double trumpet sound locators for locating aircraft. The horizontal arrangement of the trumpets (left) allowed the azimuth angle of the aircraft to be determined, while the vertical arrangement (right) allowed its height to be determined (Hunke 1935, plate 7).

Cambridge physiologist A.V. (Archibald Vivian) Hill coordinated most of the British activities in sound detection of aircraft. His Anti-Aircraft Experimental Section at the Munitions Ministry's Munitions Invention Department (MID) was known as Hill's Brigands. Hill's Brigands worked on sound detection within the larger context of setting up antiaircraft gunnery, including developing exact gunnery tables and instruments for locating, aiming at, and shooting down planes. At the MID, Hill worked with Horace Darwin, son of Charles Darwin and cofounder of the Cambridge Scientific Instrument Company, which also took over the production of the Bull sound-ranging apparatus for artillery. The experimental section's headquarters were established at Farnborough Airfield in Hampshire in 1916, and later relocated to Rochford Airfield in Essex. Work was also carried out at the National Physical Laboratory (NPL) at Teddington, the Royal Navy gunnery school HMS *Excellent* on Whale Island, Portsmouth, and the School of Electric Lighting at Stokes Bay.[33]

As in the experimental section in artillery ranging under W. L. Bragg, Hill's Brigands kept up a close collaboration with the French War Office through the Direction des inventions, and later with the Anti-Aircraft Service of the American Expeditionary Force (AEF). Whereas the British and French armies developed the Bull sound-ranging apparatus together, and the AEF then adopted it, the development of sound locators for aircraft by the different armies remained separate. During initial experiments in 1916, C. Jakeman at the NPL used the French Claude orthophone, designed for locating enemy guns in the trenches.[34] The Claude orthophone, which did not employ any trumpets, was modified to a design with four trumpets, similar to the Waetzmann sound locator. Compared with the German design, the British one employed much smaller trumpets. The first two hundred sound locators were ordered in October 1917 (Zimmerman 2001, 6), but the small trumpets proved to be not very suitable for detecting the German long-range Gotha bombers that had started to appear over the British Isles, because they emitted comparatively low frequency sound.

The French Army deployed two very different systems. One was the Baillaud paraboloid, designed by the astronomer René Baillaud: a parabolic mirror that reflected the sound from the aircraft and then directed it to the ears of the observers. The other French system was the sitemeter and telesitemeter, developed by physicist Jean Perrin. The telesitemeter was a rather large instrument with several small trumpets arranged in four circles on a plane. The complex mounting turned around the trumpets so that they faced precisely the direction of the sound. The two observers were seated on the apparatus itself and turned along with the trumpets (Bloor 2000, 201).

British and American observers concluded that the Perrin telesitemeter was the most accurate of all the listening devices, the Baillaud paraboloid precise but not nearly as precise as the telesitemeter, and the British trumpets the least accurate. The Baillaud apparatus was selective but very heavy; furthermore, Lieutenant A. Ward of the Royal Air Force evaluated it as being too complicated, requiring specialized personnel and careful training. The Perrin apparatus was less complex to operate but more difficult to manufacture, so that it could probably not be obtained in the numbers necessary. The British trumpets, though lacking in precision, were cheap, simple, and easy to transport.[35] Interestingly, British scientists used the same type of arguments against the Perrin telesitemeter as German Army representatives had used against automated sound ranging with the oscillograph system developed by Fredenhagen. Whereas for artillery ranging, British scientists and military representatives had decided that they needed the precision of the Bull automated system even though it was complicated and expensive, when it came to aircraft location, they chose not to use the more complex but more precise French instrument. The U.S. Army did not adopt the telesitemeter either. Instead, George W. Stewart of the State University of Iowa carried out experiments with long conical trumpets, even longer than those used by the German Army, to receive a better spectral distribution, including the low sound frequencies emitted by long-range bombers.[36]

Several aspects explain the British and American reservations about the French telesitemeter. One was its sheer logistics and cost. Added to that, locating aircraft with the telesitemeter—unlike the Bull sound-ranging system—was still a subjective method that relied on human observers. Hill estimated that the precision of 0.5 degrees achieved with human observers and the British trumpets was sufficient for short-range detection and for directing searchlights, suggesting that aircraft detection did not require a precision comparable to artillery ranging (Van der Kloot 2011, 405–406). All methods for sound location of aircraft that I have discussed so far were subjective methods, in which human observers detected and interpreted the sounds with their ears and located the planes using the binaural method. One of the obstacles to objective sound measurement was that the sound energies received from distant planes were rather small compared to those from heavy artillery recorded in the trenches. When Admiral Sir Henry Jackson tested some of the listening devices, he tried a microphone with the trumpet and found the intensity to be less than with the unaided ear.[37]

Despite the difficulties of recording aircraft sounds by microphones, scientists in Germany and Britain still endeavored to develop objective systems for aircraft detection. After all, the planes emitted distinctive frequencies, which were sometimes, as discussed by Waetzmann, harmonic and almost musical. The sounds originated from the engines and propellers of the planes and were specific to the model. These specific sound spectra allowed human listeners to distinguish between their own and enemy aircraft. In 1918, Fritz von Jagwitz published a report on sound vibrations of airplanes, picked up by microphone and telephone and recorded by oscillograph and string galvanometer. In the report, Jagwitz compared the specific sound spectrum of certain aircraft to the sound spectrum of musical sounds. Though the sound spectra of airplanes were not entirely harmonic, they represented a combination of several fundamental tones that often resulted in beats, and that were characteristic for each plane (Jagwitz 1918, 11–13).

The distinctive sound spectrum of particular planes drew William Tucker to the sound detection of aircraft. In 1917, Tucker had completed his successful work on hot-wire microphones for artillery ranging. The low frequency spectrum of the German G-planes, which had now started to bomb London, was unpleasantly reminiscent of the low frequencies of heavy

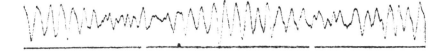

Figure 3.6
Sound recording of a German G-plane by a mirror oscillograph. The plane's two engines did not run entirely evenly, resulting in beat frequencies. The main frequencies were around 70–75 cps and its octaves. The lower line gives the time reference of 0.2 seconds between the breaks of the line (Jagwitz 1918, 13).

guns that Tucker had managed to isolate and record. He now moved on to develop similar objective methods for sound detection of aircraft, transferring practices and instruments from artillery ranging to aircraft detection. Tucker did not plan to duplicate the work of Hill's Anti-Aircraft Experimental Section, which developed devices for short-range acoustic location. Instead, he wanted to develop a system for long-range acoustic detection that would give air defense units sufficient warning time to prepare for a strike (Van der Kloot 2011, 405; Zimmerman 1997, 74). Within a few days of his proposal, the Air Inventions Board approved and established Tucker's Acoustical Research Section and its experimental grounds at Imber Court in Surrey. Tucker's initial listening apparatus comprised two parallel discs, 9 feet (2.75 m) in diameter and 7 inches (18 cm) apart, which he found gave significant magnification in combination with the microphone. A report by Hill's Brigands, comparing the discs to trumpets, found large binaural trumpets much more effective.[38] The shortcomings of the disc arrangement might explain why Tucker continued to experiment with a concrete mirror and a mirror carved out of the cliffs at Fan Hole near Dover. Both were made by the electrical engineer Thomas Mather of Imperial College and employed trumpet collectors.[39]

Tucker set out to develop frequency-sensitive microphones that were tuned to detect only the specific sound frequencies of enemy aircraft and were insensitive to other noise, including the sounds of Britain's own aircraft. He increased the sensitivity of his hot-wire microphones by using thinner wire and placed them in tube resonators and trumpets. Testing his resonating microphone system against a carbon microphone and an electrodynamic magnetophone for low frequency sounds, Tucker found his resonating microphone to be most suitable for the long-range detection of planes. Equally important for the development of sound measurement, Tucker started to use a valve amplifier with his microphones.[40] In February 1918, he spent some time with Frank Edward Smith of the NPL to work out which vacuum tubes, or "valves" in British English, gave the most satisfying results. They found the Mark II WT (wireless telephone) amplifier with BTH (British Thomson-Houston) valves, with its amplifier circuit modified, to work best. Obtaining a supply of Mark II amplifiers would pose no great difficulty, since they were in rather common use at the time.[41] In April 1918 Tucker could report that the valves had been set up in a box to be used in the field with a trumpet apparatus and microphones.[42]

On 15 January 1918, Tucker was visited by W. L. Bragg's father, William Henry Bragg, who led a group on underwater acoustics at the Board of Invention and Research. W. H. Bragg and his team had developed hydrophones for submarine warfare (Hackmann 1984, 50–55). Not unlike microphones for aircraft location, hydrophones reproduced very weak sounds from the enemy vessel, which were drowned out by the noises of the ocean and the observing ship. W. H. Bragg and his collaborators needed suitable microphones and valve amplifiers as well (Hackmann 1984, 55). According to Tucker, W. H. Bragg and Mr. Hunter from the Admiralty were much impressed by Tucker's microphones, which they considered "more faithful in its reproduction of sounds than those they have at present in use."[43] This was even more important for submarine detection than for aircraft location, since submarine sound did not seem to be characterized by one specific frequency. Six of Tucker's hot-wire microphones were selected for comparison, and Tucker speeded up production of more sensitive microphones, both for his own use and for antisubmarine work.

3.2.3 Underwater Acoustics and U-boat Warfare

At the beginning of the war, military leaders did not anticipate the role and importance that submarines would take on in naval warfare, just as trench warfare and the changing role of aircraft were not foreseen. When the German Empire developed its global colonial ambitions, Wilhelm II pushed for the rapid expansion of the German Navy in order to catch up with its British counterpart. His naval chief of staff, Grand Admiral Alfred von Tirpitz, started to build up a large modern fleet soon after his appointment in 1892. Nonetheless, at the outbreak of the war, the German fleet was not able to compete with the Royal Navy, which continued to control the oceans. As a naval strategy against a superior enemy, the German Navy turned to U-boats. In the early days of the war, submarines sank only battleships, but in 1915 they moved on to unrestricted attacks on passenger and merchant ships. Germany's escalation of the U-boat campaign was what ultimately brought the United States into the war in 1917.

The U-boat campaign channeled research and development in underwater acoustics during World War I into two main directions. One was the use of underwater sound signaling for communication between vessels. Ships could communicate visually and increasingly by wireless telegraphy, but radio signals could not be used under water.[44] Instead, underwater sound telegraphy effectively took on the role of wireless for U-boats to communicate with their own fleet. The second direction of research in underwater acoustics was the development of hydrophones to listen to and analyze the sounds produced by enemy vessels. While German efforts in underwater acoustics were largely directed at supporting U-boats in their campaigns, British efforts were antisubmarine. For the British navy, the development of hydrophones was part of a set of measures to detect, locate, and ultimately hunt down German submarines, but for the German Navy, listening hydrophones were not simply a defensive measure. German researchers developed nonresonant noise receivers (*Geräuschempfänger*) and installed them in U-boats alongside resonant signal receivers for sound telegraphy. The noise receiver allowed submerged submarines to orient themselves under water and locate enemy vessels without using their periscopes.

Lightships and other vessels had been using underwater sound signaling since around 1900, whenever fog and other weather conditions inhibited visual communication (Aigner 1922, 1–7; Hackmann 1984, 3–10). Underwater bells were driven electrically or pneumatically and could be heard several kilometers through the sea. The sound velocity was more than four times higher in water than in air. With the help of microphones placed in tanks port and starboard, operators could detect the direction of the sound by turning the ship until the signal was heard equally strongly on both sides. But when it came to developing effective systems for sound telegraphy, underwater bells were judged too slow and lacking in range. Underwater sirens were employed, but wore out quickly in the harsh seawater environment. An electromagnetic or electrodynamic sound transmitter proved a better solution. In 1912, Reginald Fessenden, a former professor of electrical engineering at Purdue University in Indiana and a pioneer of wireless technology, developed an electroacoustic sound transmitter and receiver for transmitting Morse code. His Fessenden oscillator employed an electrodynamically driven round steel plate as a membrane. In principle, it resembled a large telephone that could serve as a transmitter as well as a receiver. The Fessenden oscillator was manufactured by the Submarine Signal Company in Boston and under license by Atlas Werke in Bremen, Germany.

The German Navy's torpedo testing facility, TVK, had its own underwater telegraphy department to test underwater telegraphy systems and install them in U-boats and other vessels. After mobilization in 1914, the department lost most of its personnel, keeping the physicist H. Ament and one engineer. By 1915 it had regained its previous staffing level and had started to employ more scientists. Through the Army and Navy Radioelectric Testing Laboratory, the TVK collaborated closely with Simon's Institute for Applied Electricity at the University of Göttingen. Ament had worked in Simon's laboratory in 1906. It is no surprise that most of the scientists recruited at the TVK came from Simon's institute. Max Reich, who had headed the Radioelectric Testing Laboratory since 1909, was drafted as an artillery captain when the war started but in 1915 became scientific director of the TVK. Heinrich Barkhausen, Simon's former assistant and since 1911 professor extraordinarius of low-current electrical engineering at the Technische Hochschule Dresden, was drafted as an auxiliary scientist in March 1915. In 1916, Simon's former students Hugo Lichte and Ernst Lübcke followed. Lübcke had completed his dissertation on the recording of curves of alternating current with Simon just that year. Simon himself developed a noise receiver, for which he took leave from his professorship in Göttingen (Busse 2008, 224; Aigner 1922, 234).

Barkhausen investigated the conditions of sound propagation in seawater in the Eckernförde Bay, just outside Kiel, and in the Danzig (now Gdańsk) Bay from May to July 1915. Just as wind speed and wind direction influenced sound propagation in the atmosphere, sound propagation in the sea was influenced by water currents. Barkhausen and Lichte determined the dependence of sound propagation on water temperature, temperature gradients, and salinity. The horizontal layering of the seawater led to a downward or upward bending of the sound rays. The sound was reflected or absorbed by the seabed, depending on whether this consisted of hard rock or mud. As a result of these changing effects of sound spreading and absorption in water, the relation between the range of sound signals and the transmitted power output varied greatly. Under some conditions, absorption increased exponentially with distance, under others, it did not. This limited the reliable practical range of underwater sound telegraphy to 10 kilometers or less. Exponential absorption meant that simply increasing the sound power output would not solve the problem. The range of sound signals from the same equipment varied from as little as 2 kilometers to as much 100, due to layering and irregularities in the water. The range was usually greater in winter than in summer, when solar warming brought more temperature irregularities (Lichte 1919; Barkhausen and Lichte 1920).

The TVK was not alone in carrying out research and development in underwater acoustics in the German Baltic Sea during World War I. In 1906, the German Navy had commissioned the marine engineering company Neufeldt and Kuhnke in Kiel to develop sound transmitters for underwater telegraphy. Hans Usener, technical director of Neufeldt and Kuhnke, hired physicist Karl Heinrich Hecht for this task in 1908 (Ziehm 1988). Hecht had studied with Paul Volkmann in Königsberg (now Kaliningrad) and had previously worked at the Physikalisch-Technische Reichsanstalt and Siemens laboratories ("Dr. Heinrich" 1940; Lichte 1940). In 1911, Neufeldt and Kuhnke founded the company Signal Gesellschaft as a subsidiary to take on underwater telegraphy. In 1912, electrical engineer Walter Hahnemann joined Hecht at the Signal Gesellschaft, leading to a long-lasting scientific collaboration between the two. Hahnemann had studied at the Technische Hochschule Munich and previously worked with Telefunken, the company C. Lorenz, and the German Navy on wireless telegraphy (Zenneck 1939). Engineer and patent lawyer Alard du Bois-Reymond, son of physiologist Emil du Bois-Reymond, joined the team on underwater acoustics research and development at the Signal Gesellschaft.

From 1915, tensions began to build between the Signal Gesellschaft and the Torpedo-Versuchs-Kommando. The Signal Gesellschaft had once collaborated closely with the TVK in research and development, but this changed when the TVK stepped up its research and development of sound telegraphy systems for U-boats in 1915 and the navy established its own production facilities. The navy terminated a five-year contract with the Signal Gesellschaft and did not include the company in its research, even when the TVK used the company's test stand. From the end of 1914, the navy commissioned Atlas Werke to produce equipment jointly developed by the Signal Gesellschaft and the TVK. For the Signal Gesellschaft, Hahnemann argued that the company needed a guaranteed financial return for its research efforts to develop a genuine German system, and that Atlas Werke seemed to be producing mainly licensed technology.[45] Moreover, not only was the Fessenden oscillator, built by Atlas Werke, a licensed product, but it was licensed from a company based in the United States, which was at war with Germany by this time. Admiral Eduard von Capelle, state secretary in the Imperial Navy Office, saw things differently. From his perspective, the Signal Gesellschaft was trying to monopolize the market for underwater sound equipment, which would have disastrous consequences for the development of such equipment for the German Navy. In July 1915, the navy had evaluated the underwater telegraphy systems delivered by the Signal Gesellschaft as not yet technically mature and had emphasized the need to train personnel. It was only logical that the navy should carry out its own research and development and draw in other companies to enhance competition.[46]

Siemens and Halske embarked on their research and development in sound transmitters and receivers only during the war. In July 1914, Siemens hired physicist Hans Riegger for its physical and chemical laboratory. Riegger had completed his dissertation on wireless telegraphy under Ferdinand Braun in Strasbourg in 1910 and was assistant to Max Wien and Jonathan Zenneck at the Technische Hochschule Danzig. From September 1915, he carried out experiments in underwater sound signaling in the Flensburg Fjord, 80 kilometers north of Kiel. With Karl Boedeker and Hans Gerdien, the scientific director of Siemens Research Laboratories, Riegger developed an electromagnetically driven oscillator and a resonant signal receiver for underwater telegraphy, as well as a noise receiver. In spring 1918, Riegger and Gerdien patented an electrostatically driven membrane for sound transmission for Siemens.[47]

Despite the friction between the Signal Gesellschaft and the navy, the electromagnetic sound transmitter developed and manufactured by the Signal Gesellschaft remained standard for the German Navy's underwater telegraphy (figure 3.7). Two models were produced. Austrian physicist Franz Aigner specified the sound power output of the small model as 100–125 W, with a membrane diameter of 30 cm and a weight of 110 kg. He gave the output of the larger model as 300–400 W, with a membrane diameter of 45 cm and a weight of about 240 kg (Aigner 1922, 194). The driving frequency and resonance of both models was fixed at around 1050 cps, which was the sound frequency used by the Central Powers for underwater telegraphy. The Entente forces used a sound frequency of around 540 cps. According to Aigner, most disturbing noises of ships were in the range of 500–800 cps, which he cited as the reason for using the higher frequency (66). This

Fig. 66.

P .... Haube; Sche .... Schutzwiderstand; A .... Anker; F .... Feld; M .... Membran; D .... Deckel; De .... Drosselspule; S .... Stabrohrsystem; T .... Tisch; B .... Zapfen.

Figure 3.7
Electromagnetic sound transmitter designed by the Signal Gesellschaft, Kiel. The membrane *M* was connected to the table *T*. The table was mounted in a system *S* of four rods and tubes, driven by an electromagnet. *A* shows the anchor of the magnet. *De* shows a choke coil (Aigner 1922, 195).

choice, however, also meant that the German Navy stayed outside the broadcast frequency of the enemy, especially since the navy used an identifying sound signal to acoustically identify their own craft. According to Eberhard Rössler, the Signal Gesellschaft delivered a total of 301 underwater telegraphy systems to the German Navy (Rössler 2006, 18).

Like the Fessenden oscillator, the Signal Gesellschaft's sound transmitter could be used as a receiver as well. But it was not an efficient receiver—hydrophones with far smaller membranes attached to carbon microphones (figure 3.8) were much more sensitive.[48] As well as the signal receiver, a second type of receiver, the noise receiver, was employed on the U-boats and other vessels that needed to detect submarines, as Franz Aigner explained after the war:

While the tone receiver suffices for the normal and undisturbed recording of underwater sound signals, U-boat warfare created the need for a second type of receiver, which was mainly designed to record the noises [Geräusche] of ship propellers or alien submarines, and was therefore known as the noise receiver [Geräuschempfänger]. ... [I]ts natural resonance frequencies are strongly absorbed. This enables it to record equally both the on-board noises of a nearby ship and the working of the ship engine, pumps, electric machines, and the grinding of the propeller. All these noises have a characteristic timbre for the listener, so that with practice he will be able to

Figure 3.8
Signal receiver for 1,050 cps, on the left without a membrane. Inside, a special type
of carbon microphone was attached to the membrane. The noise receiver was similar
but had a circlip between cap and membrane for calibration (Aigner 1922, 205).

determine from the noises with absolute certainty the type of ship, the number of
propellers and their blades, and the type of ship propulsion, whether piston engine
or turbine, etc. (Aigner 1922, 203)[49]

Aigner's remarkable account not only introduced notions of signal and
noise that developed during the employment of acoustics in warfare. It also
presented a concept of "noise" that was not a nuisance to the listener at all
but, on the contrary, highly significant in terms of the information it carried about its origin. The same noise became a fatal nuisance to the crew of
the ship to be detected. This dichotomy—sound as important information
that needed to be amplified, or sound as a potentially deadly nuisance that
needed to be avoided—marked the acoustics of aircraft, artillery, and tunnels under trenches. Accordingly, the efforts of acousticians went in two
directions: noise detection and analysis on the one hand, and noise abatement on the other. While scientists tried to build more efficient hydrophones and air sound locators to detect and identify enemy vessels and
aircraft, they also tried to silence their own aircraft and ships and limit their
vulnerability to acoustic detection.

The introduction of the noise receiver meant that U-boats had to be equipped with a second set of listening devices. The navy pushed for the development of an *Einheitsempfänger*, a combined receiver for signal and noise (Aigner 1922, 205). This naturally caused design problems, given that hydrophones could be built either as resonant or as nonresonant: either hydrophone operators wanted to tune in to the frequency of a signal, or they wanted to listen to the noise and optimize their receiver accordingly.

The German Navy knew that the French and British navies were building their own listening devices to detect German U-boats. On May 1916, the submarine survey command Inspektion des U-Boot-Wesens called a meeting with Ament, Barkhausen, Lichte, and Lübcke. The background was an article in *Scientific American*, 16 October 1915, "Submarines Betrayed by Sound Waves," which discussed a characteristic U-boat tone in the high-frequency range. As reported in the article, the American electrical engineer William Dubiller had visited the French port of Cherbourg, where he had met "professor" (actually naval officer) Camille Papin Tissot and discussed the sound detection of submarines. Supposedly, no important details of sound detection had been disclosed, but the article mentioned "the introduction of a De Forest audion amplifier between the apparatus and the telephone receivers, ultimately resulting in a range of 55 miles" for sound detection ("Submarines Betrayed" 1915, 333). One incident lent credibility to the claims made in the *Scientific American* article: an enemy submarine had apparently followed the German U-boat *U 45* for a long distance, presumably using acoustic detection.

The TVK investigated the sounds of the U-boats carefully but could not detect any specific U-boat tone. The sounds of a submerged U-boat consisted of its propeller noise, electric engine noises, and the grinding of the shafts. While hydrophone listeners could identify propeller noise at distances of 2–3 nautical miles after considerable practice, other sounds could be heard only at maximum distances between 1,000 and 2,000 meters.[50] The Inspektion des Torpedowesens concluded that there was no specific U-boat tone but considered that the British navy might have superior noise detection instruments. The German U-boats could only use their noise receivers when resting on the seabed or traveling at very slow speed. The British patrol boats seemed to employ hydrophone equipment that could be used at cruising speed. If possible, the Inspektion recommended, enemy vessels should be searched for hydrophones before destruction and the apparatus handed over for study.[51] The Inspektion des U-Boot-Wesens ordered pumps and transformers to be silenced by mounting them on rubber pads. Harmonic sounds and periodic noises, especially, were to be avoided.[52]

The loudest and most characteristic sound that would give away the presence of a U-boat remained the propeller noise. Since March 1915, Ludwig Prandtl, at the Modellversuchsanstalt (Model Research Laboratory) of the University of Göttingen, had participated in researching sound propagation under water. This was part of Simon's development of a noise receiver at the Institute for Applied Electricity, which also involved Emil Wiechert at the university's Geophysical Institute (Schmaltz 2013, 4–5). In May 1917, Prandtl proposed studying the noise produced by submarine propellers and developing quieter ones (Schmaltz 2013, 9). Soon thereafter, he carried out observations on a U-boat test voyage at Travemünde, confirming that cavitation was the source of the noise. In December 1917, the Inspektion des U-Boot-Wesens commissioned Prandtl to design and test new propellers. Prandtl and his collaborator at the Model Research Laboratory, Albert Betz, executed calculations and model experiments in the wind tunnel and mapped the noise patterns of propellers. The new propellers subsequently designed by Betz and Prandtl produced less noise in tests— but they also proved to be less efficient, reducing the velocity of the U-boat. The loss of speed was not acceptable to the navy. Further effort was now devoted to designing propellers that produced less cavitation without compromising their efficiency in propelling the boat. The Kiel Mutiny and the termination of World War I put an end to the project before testing of the silent propellers was completed.[53]

The discussions about U-boat noise, the development of noise receivers, and the possible superiority of British listening apparatus all emphasized the importance of binaural direction sensing. Between February and August 1917, Max Wertheimer traveled to Kiel and Travemünde eight times to participate in test voyages and to adapt binaural direction sensing to underwater conditions. Binaural direction equipment was used only on patrol boats; testing for its use in U-boats was not completed before the war ended (C. Hoffmann 1994, 269–271; Aigner 1922, 244–250).

In 1918, the submarine threat had turned around. The Allies sank more German U-boats, and Allied submarines increased their activities in the

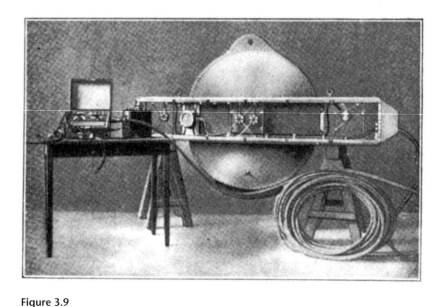

Figure 3.9
Setup of hydrophones to be towed by a vessel for direction sensing by the German Navy. The arrangement allowed observations that were less disturbed by the sounds from the patrol ship. In this photo, the canvas strung around the wooden frame has been removed, and four hydrophones can be seen. The large disc in the center produced an acoustic shadow to determine the direction of the sound. Another pair of hydrophones was attached on the other side of the disc (Aigner 1922, 241).

North Sea. The German Navy set up a listening station off the island of Heligoland in the German Bight. Listening stations were also set up at the English Channel off the Belgian coast to monitor Allied naval traffic after the Ostend and Zeebrugge raids. These stations became operational in July 1918. One sound-ranging section of the APK took part in this project, as the measurement principle was analogous to the objective sound ranging that had just been introduced for artillery. The microphones were tuned to the frequencies of ship propellers, placed on tripods on the seabed, and connected to the receiving station by sea cable. The readings of the microphones were then recorded by an oscillograph or analyzed by human observers using telephones.[54]

The historiography of underwater acoustics in Britain and France during World War I has so far focused on tracing the history of sonar (in Britain

called asdic) and contrasting passive listening devices with the development of active echo ranging, which employed ultrasonic piezoelectric
transducers.[55] In Britain, most research and development on underwater
acoustics was divided between the experimental station at Hawkcraig
(under the Royal Navy's Anti-Submarine Division) and the Board of Invention and Research, run by civilian scientists. Work on underwater telegraphy was carried out at the Torpedo School and transferred to the Signal
School in 1917 (Hackmann 1984, 21). The Hawkcraig station was established by Captain Cyril Percy Ryan. He began his work on detecting U-boats
acoustically in 1914 after leaving the navy for three years to work on wireless with the Marconi Company (Hackmann 1984, 21–23). For the Board of
Invention and Research, Ernest Rutherford started to work on underwater
acoustics and submarine detection at the University of Manchester in summer 1915, with his colleagues Albert Beaumont Wood and Harold Gerrard.
Wood and Gerrard moved to the Hawkcraig station in November 1915,
where William Henry Bragg joined them as director of research in April
1916. The relationship between Ryan and the civilian scientists was tense
from the outset and only grew worse. Bragg and his colleagues had problems obtaining instrument fitters and access to the workshop, and they
finally left Hawkcraig at the end of 1916. A new listening school opened in
August 1917 at HMS *Sarepta*, at Weymouth in Dorset, and was transferred to
nearby Portland in April 1918 (Hackmann 1984, 25).

As in the German case, industry became involved in the development of
hydrophones, with companies including British Thomson-Houston (BTH)
and Western Electric. BTH was also the producer of the amplifier valves
that Tucker used for aircraft detection with the Mark II wireless telephone
amplifier. The U.S.-based company Western Electric abandoned its public
claim of neutrality after 1916 and entered war-related research in underwater acoustics and artillery ranging, working on valve amplifiers as well.
George Nash, who had been chief engineer at the London branch of
Western Electric since 1911, developed a hydrophone referred to as the
"fish," which was towed behind a patrol boat.[56] As the German Navy had
realized, the Allies were developing listening devices that could be used
while the boat was patrolling. By towing the hydrophone at a sufficient
distance from the observing boat, skilled observers could distinguish
between the sounds emanating from the patrol boat and those emanating
from the U-boat to be observed (Hackmann 1984, 60).

The British scientists collaborated very closely with their French and American colleagues in underwater acoustics research and the development of hydrophones, though initial frictions, originating in the British Admiralty, had to be overcome. One of the most important contributions to the British research was the superior high-frequency valve amplifier that the French made available in 1916 and 1917 (Hackmann 1984, 55; Katzir 2012, 150). American interest in hydrophones started with the U.S. entry into the war in February 1917. The U.S. Naval Consulting Board had been established in 1915 at the suggestion and under the leadership of Thomas Edison. The Submarine Signal Company, General Electric, and Western Electric pooled their resources at the Antisubmarine Laboratory that opened in Nahant, Massachusetts, in April 1917. Academic physicists were excluded from the Nahant facility, which prompted the National Research Council to establish a second experimental station at the submarine base in New London, Connecticut.[57] Meetings between British, French, and U.S. scientists took place in June and July 1917 (Hackmann 1984, 39–40). American hydrophones were installed on both U.S. and British ships. Willem Hackmann describes British hydrophones as designed to be resonant to specific frequencies, whereas U.S. hydrophones were mostly nonresonant. The American scientists experimented with placing the microphones in rubber rather than wooden casing, a material that Ryan had adapted at Hawkcraig for a distinctively small, streamlined hydrophone known as the "rubber eel." Ryan used a Bell microphone with a Marconi valve amplifier (Hackmann 1984, 61).

There is no indication that the German Navy ever succeeded in capturing acoustic listening gear for U-boat detection from the Allied navies. The British navy, in contrast, did manage to examine the acoustic gear of two captured German U-boats, *UC 5* and *UG 44*, in 1916 and 1917. The British realized that the U-boats could use their listening gear only when their engines were off, and therefore the British considered their own listening devices superior to the German ones.[58] But aside from competing claims to superiority, underwater acoustics—previously left to technology entrepreneurs rather than scientists—became a major research field in Britain, France, Germany, and the United States during the war. In artillery ranging and aircraft detection, microphones still competed with the human ear aided only by collecting trumpets. In underwater acoustics, no such competition existed. Underwater sounds were almost always communicated through electroacoustic hydrophones for sound measurement.

3.3   The Transformation of Sound Measurement

The military use of acoustics during World War I had fundamentally altered the methods, practices, and instruments of acoustical measurement. Before the war, acousticians had listened to and analyzed musical sounds; they now had to deal with complex soundscapes where machine sounds, which were often harmonic and periodic, were sometimes mixed with diffuse noises. Out of this complexity, certain sound objects (the firing of a particular gun, the distant sounds of an enemy ship) had to be singled out and investigated. The specific surveillance interest at stake defined which sound objects were interesting, which disturbing.

The human ear, aided by microphones, hydrophones, or trumpets, remained the most powerful acoustic instrument for analyzing the composition of complex sound objects, but for precise registration of sound events, human observers were inferior to objective measurement by automated systems, largely because of the delayed and uneven response time of the observers. Automated systems also produced recordings that could be reconsulted later if doubts arose. Instead of recording sound signals on a phonograph, fast-reacting string galvanometers and oscillographs picked up the weak signals of microphones and recorded them on photographic paper.

The challenge of automated objective measurement was to develop microphones that responded to the sound objects to be investigated without reacting to other, disturbing noises. In other words, the microphone or the registration apparatus had to act as a filter—filtering out certain sound frequencies while allowing others to pass. The microphones that were initially used for sound ranging were conventional telephone receivers. These were not designed to detect and single out the low-frequency sounds originating from firing artillery. If at all, they were optimized for speech transmission on telephone lines and detected all kinds of noises. William Tucker succeeded in building a hot-wire microphone that registered only the pressure waves of large artillery guns and ignored all other sounds, including the shell wave originating from high-velocity gun shells traveling at supersonic speed (Winterbotham 1918, 34; Tucker and Paris 1921; MacLeod 2000; Van der Kloot 2005, 276–278).

The British automated system for sound ranging exemplifies many of the transformations in sound measurement that took place during the war. The battlefields of World War I changed acousticians' view of what signal had to be detected and what noise had to be suppressed or ignored in order to measure that signal. Furthermore, scientists had developed objective instrumental systems, such as the Tucker hot-wire microphone arrangement, that could filter out this signal from all the noise. The question of what was signal and what was noise did not, however, become any more objective; it remained subject to the observer's intentions.

Tucker was probably the most successful in building microphones that were sensitive to certain frequencies while filtering out others, but others tried as well. When the German Army replaced its subjective sound-ranging squad with an objective instrumental system toward the end of the war, it failed to develop a microphone similar to Tucker's. Angerer and Ladenburg reported on experiments, however, undertaken in July 1916, that used pressure wave meters instead of microphones. These were thin strips of paper that were ripped apart by the air pressure, triggering the measurement circuit (Ladenburg and Angerer 1918, 18). Tucker's hot wires suspended over box resonators and Angerer and Ladenburg's paper strips were both surprisingly simple devices, based on simple physical principles. That the German scientists did not manage to come up with a device as successful as Tucker's indicates, however, that these objects and principles though simple were far from trivial.

The hydrophones that German, British, and American scientists developed for underwater acoustics were either frequency-specific or responded to a broad range of frequencies, depending on their design. While the British and American scientists distinguished between resonant and nonresonant hydrophones, German scientists distinguished between signal receivers and noise receivers. The basic characteristics of these two classifications were the same. The German resonant hydrophones, or tone receivers, were optimized for the signaling frequency of 1,050 cycles and were mainly used for underwater sound signaling (Aigner 1922, 66, 205). The nonresonant hydrophones, or noise receivers, were used for listening to other vessels (203).

The work of Erich Waetzmann shows how far acousticians had moved from musical listening to valve amplifiers and automatic registration by the end of the war. In 1921, Waetzmann published a two-part paper on his research into sound monitoring, sound analysis, and sound location of aircraft during World War I. At this stage, aircraft detection still relied largely on human observers listening with funnels or the unaided ear rather than microphones. Waetzmann named two fields of practical application for researching aircraft sound. First, to build apparatus for sound location of aircraft, the kind of noise to be located must be known. And second, to build a "soundless" aircraft, knowledge of the sounds was necessary; otherwise they could not be eliminated systematically.[59]

Both Waetzmann and Jagwitz had used a language of musical acoustics and musical listening, trying to compare the sounds originating from aircraft to the sounds of musical instruments. Waetzmann, though, also acknowledged that analyzing aircraft sound was a tricky and complex affair, not really comparable to the analysis of musical sounds.[60] It was often not possible to determine the frequency of the main tone, or even to assess the components of the combined tones. Using tuning forks and pitch pipes, until then the backbone of experimental investigation in acoustics, was "completely inadequate" (Waetzmann 1921a, 166). The use of Helmholtz resonators required a large amount of skill and practice in listening, especially when the plane was up in the sky. While certain sound sources, such as the engines and the propellers, could be singled out, it was the aircraft as a whole that mattered. An engine and a propeller sounded very different depending on whether they were placed in a test stand or mounted in an aircraft, and whether the aircraft was on the ground or in the air.

Most of Waetzmann's investigations were subjective observations that he and his assistants, including Wilhelm Moser, had made with their unaided ears. But the researchers used a carbon microphone designed by Philipp Brömser to compare a model propeller mounted on an electric motor with the sound impression of an airplane at a distance of one kilometer.[61] For the former recording, Waetzmann's group used a valve amplifier and an oscillograph, for the latter, the string galvanometer designed by Angerer and Karl Wolff. Commenting on Waetzmann's article, Ludwig Prandtl reported his own experiments on aircraft propellers at the Aerodynamic Testing Laboratory in Göttingen, some of which were carried out in a wind tunnel (Prandtl 1921). While the acoustic aspects were probably not his main concern during the propeller experiments, Prandtl had nevertheless paid considerable attention to them. This should come as no surprise, since Prandtl and his collaborators also conducted acoustic tests on another type of propeller: in 1917, the acoustic tests of the U-boat propellers were not performed in a cavitation tunnel, as one might expect, but in a wind tunnel of the Göttingen Aerodynamic Testing Laboratory (Schmaltz 2013, 12).

There were other links between acoustic research for aircraft detection and the acoustic research carried out at the Torpedo-Versuchs-Kommando in Kiel. In 1923, Ernst Lübcke described experiments on aircraft detection undertaken by the TVK during the war (Lübcke 1923). As the Allied forces began to deploy planes against submarines, the TVK started to develop an acoustic method for detecting aircraft from the submerged U-boat. Large funnels could not be used, since they would alert the enemy to the submarine's presence, but the periscope was small enough to go undetected. Because it was difficult to spot aircraft visually through the periscope, the TVK intended to attach a small microphone to it. According to Lübcke, experiments were carried out only on land, and the listening apparatus was never mounted in an actual U-boat. The initial experiments had shown that the sound of seaplanes could not be heard with sufficient certainty over longer distances.

Lübcke's account nevertheless shows how significantly acoustic measurement practices had changed in the course of the war. Like Waetzmann, the TVK had used carbon microphones constructed by Brömser. Lübcke specified a two-valve amplifier equipped with double grid valves, made by Siemens and Halske, to amplify the electric signals from the microphone. The double grid valves were probably SSI tetrodes based on a patent by Walter Schottky in 1916, which Siemens started to produce in 1917. Brömser's carbon microphone did not work well because of the microphone noise it produced. Though the valve amplifier allowed almost unlimited amplification in theory, in practice, every sound made by the aircraft was drowned in the noise produced by the microphone itself. Telephone receivers, which were based on electromagnetic principles, did not generate the same kind of noise, and they produced satisfactory results in combination with the valve amplifier. The telephone and amplifier reproduced higher frequencies better than they did lower, resulting in a distorted representation of the sound impression, but it was still possible to identify the characteristic sound of the aircraft (Lübcke 1923, 100).

Lübcke's account of microphones and amplifiers matched that of Franz Aigner in his *Unterwasserschalltechnik* (Underwater sound technology) of 1922. Valve amplifiers had become a crucial element of acoustic measurement practices during the war, whether in Germany or in France and Britain. Suitable two-valve amplifiers had been available to both sides since 1917. In theory, valve amplifiers could amplify the sensitivity of sound detectors unrestrictedly. For practical purposes, however, Aigner named an amplification factor of 500 as the limit. The sound signal to be amplified was accompanied, and finally drowned out, by all kinds of noises. Some of these disturbing noises originated from the environment, such as the noises of the ship and the ocean in underwater acoustics, or the wind in aircraft detection. Others, though, originated in the recording and amplifying circuit itself. Carbon microphones emitted a specific microphone noise that came from the carbon contacts. Electromagnetic telephone receivers did not produce this noise, but they were less sensitive (Aigner 1922, 80).

Microphones were not the only source of noise in electric circuits. In 1915, Heinrich Barkhausen had moved from his initial research on underwater acoustics to study the characteristics of thermionic valves and their behavior in circuits (Wein 2011, 67–74; Lunze 1981, 20–21). In 1917, he filed a patent on how to reduce disturbing noises in an amplifying telephone circuit.[62] To be sure, the driving force behind thermionic valve development was not acoustic amplification but wireless telegraphy and telephony. Barkhausen's interest in valves originated in his idea of using electromagnetic waves rather than sound for underwater telegraphy. Experiments by the TVK and theoretical calculations by Arnold Sommerfeld, however, showed that electromagnetic waves could not be employed because they were too rapidly absorbed in seawater.[63]

3.4   Wireless and the Rise of Electroacoustics

Before World War I, the rapid developments of wireless telegraphy and acoustics research were only weakly linked, but during the war, the two fields grew increasingly close, for several reasons. One was the growing importance of electrical measurement technology and electrical amplification in acoustics, discussed in chapter 2. Another was the study of wave propagation in the atmosphere and under water for both electric and acoustic waves. While electric and acoustic waves were distinct phenomena and behaved differently in the two media, there were many parallels, opening up a whole range of analogies. These allowed scientists and engineers to understand the propagation of one type of wave by looking at the analogous phenomena in the other. Mathematical formalism was easily translated from the electromagnetic wave to the acoustic wave and back. The third important aspect of the convergence between acoustics and wireless was that many scientists and engineers involved in sound location and underwater telegraphy had backgrounds in wireless and not in acoustics research.[64]

Wireless had a military history in Germany that preceded World War I by more than a decade. Ferdinand Braun had worked with Siemens and Halske on a wireless system for the German Army. Adolf Slaby, professor of electrical engineering at the Technische Hochschule Berlin-Charlottenburg, and his assistant, Georg Graf von Arco, worked with the Allgemeine Electricitäts-Gesellschaft (AEG) on a system for the German Navy. In 1903, Kaiser Wilhelm II pushed for the formation of Telefunken (in full: Gesellschaft für drahtlose Telegraphie System Telefunken) as a joint venture between Siemens and Halske and AEG (Hars 1999, 153–180). The desire to have a single system and one producer of wireless technology in Germany was driven by both military and economic concerns: the German Empire wanted the various branches of the German military to use a unified system, which had to be able to compete with the monopoly and patents of the British company Marconi.

Simon's Radioelectric Testing Laboratory closed down at the beginning of the war. Instead, the Technische Abteilung Funk (Technical Division of Wireless, sometimes abbreviated as Tafunk), headed by Max Wien, organized German military research and development on wireless during World War I (Wagner 1937, 66). Wien had invented the Telefunken wireless transmitter known as the Löschfunkensender and was one of Germany's leading experts in the field. In the course of the war, valve transmitters took over from other wireless transmitter technologies, including spark transmitters, arc transmitters, and high-frequency alternators. This led to intensive research on and development of vacuum tubes and amplifier circuits, as well as their mass production. Whereas the early spark transmitters broadcasted over a large range of frequencies, interfering with all other wireless transmitters in the range, vacuum tube transmitters permitted use of a small frequency band only, making it possible to broadcast several wireless stations on distinct frequencies. Even before wireless telephony, wireless operators used telephone receivers in wireless telegraphy instead of, or in combination with, Morse printers. The Morse code simply produced clicks in the telephone receiver, which were then transcribed by the operator.[65]

Although wireless telephony had been possible with prewar continuous arc transmitters, its acoustic quality improved considerably with the new valve transmitters. The success of valve transmitters and the possibilities for amplifying sound electroacoustically through valve amplifiers meant that the electric arc, studied so systematically by Simon and his students before the war, lost its relevance as a technological object. The sound-producing and sound-consuming properties of the electric arc, fascinating as they might be, were now no more than a curiosity. Much smaller and lighter valve transmitters facilitated wireless telephony in aircraft and portable wireless sets in the trenches. During the war, many young men encountered using wireless technology or were trained in its methods.

When Walter Hahnemann and Karl Heinrich Hecht developed underwater sound telegraphy at the Signal Gesellschaft, their work was carried out in almost complete analogy to wireless telegraphy. Hahnemann and Hecht published a paper in the *Physikalische Zeitschrift* of 1916, introducing the concepts of the *Schallfeld* (sound field) and *Schallantenne* (sound antenne) in analogy to concepts of electromagnetic fields and wireless antenna (Hahnemann and Hecht 1916). They took Rayleigh's *Theory of Sound* as a point of departure to develop their theory of the sound field. Rayleigh's book had been important in understanding electromagnetic waves and the development of electromagnetic field theory. But Rayleigh himself had never used the concept or the analogy of a sound field. Hahnemann again explicitly applied and extended the analogy between wireless telegraphy and sound telegraphy in a paper published with Hugo Lichte in *Die Naturwissenschaften* in 1920, which described sound transducers as *Schallsender* (sound transmitters) and hydrophones as *Schallempfänger* (sound receivers), and compared the dimensioning of a sound antenna to that of an antenna for electromagnetic waves (Hahnemann and Lichte 1920, 873–875). Franz Aigner adopted Hahnemann and Hecht's understanding of sound propagation in terms of the sound field in his *Unterwasserschalltechnik* of 1922, and he extended the analogy to the concepts of the *Strahlungsenergie* (radiant energy) and *Sendeleistung* (transmitted power output) of a sound transmitter (Aigner 1922, 81). In the handbooks of the interwar period, the sound field had become part of the conceptual understanding of sound propagation far beyond underwater acoustics alone.[66] The introduction of the sound field was a key innovation in acoustics, enabling researchers to conceptualize space and the propagation of sound within it in a profoundly new way.

Underwater sound telegraphy became important to the theory and design of large loudspeakers as well. One of the challenges of sound telegraphy was to produce efficient transducers that could propagate sound output of several hundred watts into the water. These underwater sound transmitters were basically rather large and impressive electromagnetic or electrodynamic loudspeakers that operated at only one frequency. In 1919, Hahnemann and Hecht published a series of papers about theoretical properties of sound transmitters and sound receivers as a generalized theory of electroacoustic transmitters.[67] The background for their theory was, of course, their development of the Signal Gesellschaft's sound transducers. To develop the theory of the electromechanical transformer, Hahnemann and Hecht drew on the well-understood theory of the electromagnetic transformer, which Steinmetz had explained in 1897 using equivalent circuit diagrams.[68] Hahnemann and Hecht's approach was to substitute the mechanical movement of the coil driving the membrane of a sound transmitter with a secondary circuit, which was equivalent to the mechanical system. Instead of driving a coil and membrane, the primary circuit would induce an equivalent oscillation in the secondary circuit (figure 3.10).

Hahnemann and Hecht thereby replaced a coupled system, containing both electrical and mechanical components, with a purely electrical representation of the system. The electrical representation of coupled systems became the cornerstone of the theory of electroacoustic transducers and the use of equivalent circuit diagrams, which would become prominent in the interwar period. The sound field, the equivalent circuit diagram, and electroacoustic technologies introduced a new language into acoustics. As acousticians had previously used musical terms to describe sound and

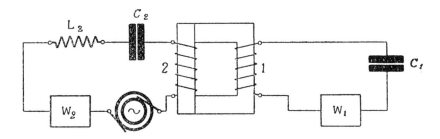

Figure 3.10
Electric circuit diagram of an electromagnetic transducer as a sound receiver. The circuit of the transducer was equivalent to that of an electromagnetic transformer. The circuit 2 represents the mechanical movement of the speaker, with L_2 and C_2 being equivalent to the elasticity of the membrane, the mass of the oscillating parts, and the mechanical and acoustic radiation resistance (Hahnemann and Hecht 1919b, 245).

sound technologies, the language of electroacoustics was derived from electrical engineering.

3.5 The Great War and the Transformation of Acoustics

Historians writing about sound location during World War I have so far mainly focused on the military context, discussing the relevance of technologies to the outcome of certain battles and the role they played in a long-term development of military technology (Hackmann 1984; Hartcup 1988; MacLeod 2000; Zimmerman 2001). From the perspective of the history of science, I argue, we should shift the debate toward the ways in which the employment of acoustics in the war changed acoustics as a discipline.[69] The acoustic detection of aircraft, for example, came late in World War I and did not have a significant effect on its outcome. Similarly, the success of hydrophones in detecting and destroying U-boats has been judged rather limited.[70] For the development of acoustics as a science, in contrast, their contribution was quite important.

There are considerable differences between the German, British, and French histories of wartime acoustics. More significant, however, are the similarities in the methods that all the belligerents employed. These similarities arose partly from the fact that the international physics community had developed and shared its methods before the war, even though these methods had not necessarily been used in acoustics. Measuring very small currents with string galvanometers, for example, was a standard practice in precision measurement, which scientists then adapted to acoustic measurement.[71] On the other hand, the different parties also carefully monitored, analyzed, and reacted to their enemies' methods and instruments of acoustic surveillance. As a result, methods of acoustic detection and counter-methods of acoustic disguise and noise abatement were coproduced.

I have so far discussed the transformations of sound measurement, the prominence of electroacoustic technologies, electric sound amplification, the relationship between acoustics and wireless, and the use of electrical analogies and equivalent circuit diagrams, all of which shaped the electroacoustic agenda that has characterized the practices and understanding of acoustics ever since. What had changed fundamentally by the end of the war was not only how acousticians studied sound, but also the sounds that they studied and the ways in which they characterized them. World War I prompted scientists to study all kinds of noises: aircraft and U-boat noises originating from propellers, engine noises, aerodynamic noises, noises created by pumps and other kinds of machinery, the cacophony in the trenches, sounds created by the wind and the ocean, and not least the noises emanating from electric circuits. What scientists regarded and investigated as a signal, and what they regarded as a noise, depended heavily on the context of the investigation. The microphones used in the various systems of sound detection were not impartial agents that registered every sound indiscriminately. Microphones were sensitive to certain frequencies or frequency bands. They needed to be shielded from wind and atmospheric disturbances. The scientists and engineers who designed and used the microphones had to know which sounds to amplify and which to suppress—and the noise that needed to be suppressed and filtered out in one kind of measurement might be the important sound that needed to be amplified and analyzed in another.

The efforts to silence machinery during World War I constituted the first systematic engagement of scientists with the task of avoiding unwanted sounds, usually known as noise abatement. Of course, unwanted noises were not a new phenomenon in society. Concerned and annoyed citizens had founded noise abatement societies in Europe and North America from 1900 on, such as the Society for the Suppression of Unnecessary Noise in

New York in 1906 and the Deutscher Lärmschutzverband (German Association for Protection from Noise) in 1908. In these societies, musicians, writers, artists, doctors, and other members of the urban bourgeoisie complained about the rising noise levels of the industrial city (E. Thompson 2002, 115–130; Bijsterveld 2008; Braun 1998).

In these noise abatement committees, acousticians were conspicuous by their absence—before 1914, physicists and engineers had written hardly anything on the problem of noise abatement.[72] Hans-Joachim Braun has argued that work on noise abatement came to a halt during World War I (Braun 1998, 253). This may be true for citizens' noise abatement committees in the urban centers, but regarding scientists and engineers, I would argue the opposite: physicists' attention was drawn to noise abatement precisely by the roar of the battlefields, and not by the calls of hygienists or the noisy streets of Berlin or London. While scientists studied the sounds produced by the enemy, they simultaneously tried to silence their own machinery. The German Navy put considerable effort into investigating and deadening every sound that could give away the position of its U-boats. British and American scientists involved in U-boat and aircraft location also worked on avoiding noise. A soundproof headpiece protected hydrophone operators from the noises of their own patrol vessel, and sound-locating trumpets had to be shielded from local sounds, especially from the noise produced by the searchlight engine.[73] In fact, Waetzmann named noise abatement as one of his main reasons to study aircraft sounds—because knowledge of sounds was a prerequisite for the building of a "soundless aircraft" (Waetzmann 1921a, 166). Silencing aircraft was, at least in its origins, not a civic demand but a military necessity.

The deployment of acoustics and acousticians in World War I laid the foundations for scientists' engagement with industrial noises and industrial noise abatement. It also laid the foundations for industrial psychology, the pioneers of which included Hans Rupp and Frederic Charles Bartlett. Rupp was appointed professor extraordinarius of industrial psychology at the University of Berlin in 1921 and took on the Department of Applied Psychology the year after. He established a direct connection between testing candidates for artillery ranging in the war and methodologies and practices in industrial psychology (C. Hoffmann 1994, 275, 277; Rupp 1921a, 1921b).

Candidates for listening posts for artillery ranging had to be tested for their response time, for their ability to separate the low-frequency sounds of heavy artillery from the din of the battlefield, and possibly for their sound direction sense. Listening posts for aircraft location and hydrophone operators, in contrast, required candidates to learn to listen to very weak sounds and to separate machine sounds from environmental sounds originating from wind, ocean currents, and sea animals. Emily Mary Smith and Bartlett of the Cambridge Psychological Laboratory developed aural tests during the war to select hydrophone operators for submarine detection (Hackmann 1984, 22). Bartlett, who was deputy director of the laboratory, then set up a unit to test personnel for hydrophone work at Crystal Palace in London. Smith and Bartlett published their experiments on the audio acuity of sounds of weak intensity in 1919 and 1920.[74]

As Waetzmann had observed for analyzing the sounds of aircraft, Smith and Bartlett found the use of traditional acoustical instruments to be inadequate to the new task. They started out using a Politzer acoumeter to measure the acuteness of hearing but had to conclude that the results from listening in free air could not be compared directly to those from listening to sounds conveyed over a telephone—the task of the hydrophone operators (Smith and Bartlett 1919, 103–104). Instead of picking up sounds by microphones or telephones, the Cambridge psychologists produced test sounds electrically, using circuits with buzzers and a tuning fork interrupter.[75] Smith and Bartlett's testing of hydrophone operators forms part of the larger story of experimental psychology that unfolded with industrial warfare. In his *Psychology and the Soldier* of 1927, Bartlett explained the psychologist's role in the technological systems of warfare:

A modern army is an immense and complicated organisation demanding highly specialised skill and knowledge in many of its branches. In recent years it has advanced towards mechanisation in all directions. Yet it still depends ultimately for its success upon the degree of insight and understanding with which the men who compose it are treated. Mechanical transport, modern artillery, tanks, aeroplanes, wireless apparatus, gas warfare, technical methods for the detection of enemy guns, aeroplanes, and submarines—in fact all the numerous applications of physics, chemistry, and engineering are at the mercy of the human mechanism by which they are employed. It is the task of the psychologist to attempt to understand that mechanism. (1–2)

Bartlett, who was appointed to the Industrial Fatigue Research Board after the war, left no doubt about the connection between military psychology and industrial psychology:

A modern army is, in fact, both for complexity and for specialised activity compa-rable with modern industrial organisation. In both it is common to meet individuals who are set to work for which their physical and mental aptitudes thoroughly unfit them. Such individuals inevitably waste time, money, temper, and are an obstacle to the maintenance of discipline and the development of morale. Many of the resulting difficulties could perfectly easily be overcome by the use of psychological tests for special abilities. (38)

In this view, the driving force behind investigating the effect of noise on the industrial worker was not the protection of workers against damaging industrial noise, but the fact that the industrial world was inevitably "at the mercy of the human mechanism."

Scientific activities were demilitarized after the war in both Germany and Britain. Many of the scientists involved did not stay with an acoustics-related research agenda but moved back to subjects such as radioactivity, atomic physics, and quantum theory. Among these were the Braggs (father and son), Ernest Rutherford, and Max Born. Others, however, continued to work on the new acoustics and extended its practices from World War I into the interwar period. Scientists published research that they had carried out during the war, and a large number of patent applications that indi-viduals and companies had filed and kept secret during the war were now made public. Although Britain experienced a demilitarization of scientific activities, it had a greater continuity of military establishments between the war and the interwar period than Germany did. The unbroken existence of institutions and programs allowed William Tucker, Albert Beaumont Wood, and others to continue their military research agenda more or less seamlessly.

The situation of their German colleagues was different. Germany's demilitarization was enforced by the Versailles Treaty, which did not allow the Weimar Republic to maintain an air force or a submarine fleet, and allowed only a small army, the Reichswehr. Despite the many actors from the scientific establishment who lobbied for military and scientific collabo-ration to continue after the war, and the German military's frequent secret violations of the treaty's terms, the market for military research and devel-opment in Germany between 1919 and the 1930s was in fact very small.[76] Many physicists returned to their academic positions; others found employ-ment in industry, to which they brought research and development struc-tures and practices from the military-scientific-industrial collaboration during the war. German industry—companies like Siemens and Halske,

AEG, the Atlas Werke, and the Signal Gesellschaft—also had to demilitarize and adapt its products to a peacetime market. Hugo Lichte moved from the Torpedo-Versuchs-Kommando to the Signal Gesellschaft in 1919, and in 1920 his former colleague Ernst Lübcke moved to the Atlas Werke, the Signal Gesellschaft's competitor. Lichte's 1919 paper on the acoustic layering of the sea became foundational for acoustic oceanography (Wille 2005, 15). Lichte himself, however, left ocean acoustics when the Signal Gesellschaft got into financial difficulties and was finally liquidated. He moved on to the AEG research laboratories, where he headed the development of the German sound motion picture system (Mielert 1985).

4 Between Science and Engineering, Academia and Industry: Acoustics in the Weimar Republic

4.1 Acoustics between Science and Engineering

World War I had reconfigured acoustics as a scientific field. Wartime activities and practices were continued into peacetime institutions and projects. To find out what acoustics looked like in Germany during the interwar period, we take a look at two extensive handbooks of the time. Ferdinand Trendelenburg, physicist at Siemens Research Laboratories in Berlin, edited *Akustik*, which appeared as volume 8 of the *Handbuch der Physik* in 1927. In 1934, two volumes of *Technische Akustik,* edited by Erich Waetzmann, professor of physics at the Technische Hochschule Breslau, were published in the monumental *Handbuch der Experimentalphysik.*[1] Both handbooks gave a comprehensive view of the new areas into which the science of acoustics had expanded in the previous decades. This acoustics of the interwar period showed surprisingly little resemblance to the works of Hermann Helmholtz and Lord Rayleigh, which had formed the cornerstones of scientists' engagement with sound in the late nineteenth and early twentieth century. By the late 1920s, the realm of acoustics had incorporated many new technical applications and, in fact, new sounds and soundscapes. Under the heading of electroacoustics, acoustics had become closely intertwined with media technologies, electrical engineering practice, and electroanalog thinking. The two books contained chapters on loudspeakers, sound motion pictures, magnetic recording, long-distance transmission of sound on wires, and radio broadcast. The subjects of architectural acoustics and noise abatement had also been added to the acousticians' agenda since the turn of the century.

Most of the authors of *Akustik* and *Technische Akustik* were located not at universities but at the German institutes of technology (Technische

Hochschulen) or in industrial research laboratories. Several of the *Technische Akustik* authors worked at the Heinrich Hertz Institute for Oscillation Research, which was co-sponsored by the Technische Hochschule Berlin-Charlottenburg, government institutions, and industry. Most the authors in both handbooks were based in Berlin, which was home to most of Germany's electrical industry and central government institutions.

The acoustics of musical sounds and musical instruments, which thirty years earlier had been at the core of acoustics research and its applications, had now become a side story. In Trendelenburg's *Akustik*, Erich Moritz von Hornbostel wrote the chapter on pitch systems in tonal music, while C. V. Raman, professor of physics at the University of Calcutta, wrote about musical instruments. Raman had worked on the acoustics of classical musical instruments from India and Europe before he turned to spectroscopy, which earned him the Nobel Prize in Physics in 1930 (Raman 1988). He was the only *Akustik* handbook author not based in Germany. Trendelenburg himself authored the chapter on speech sounds, while Erwin Meyer wrote on human hearing. Meyer was a former student of Waetzmann's and had worked at the Telegraphentechnisches Reichsamt (German Telegraph Technology Office) before he became head of the acoustics department of the Heinrich Hertz Institute in 1929. Trendelenburg and Meyer were the most prominent figures of a new generation of acousticians. Both had participated in the war as soldiers and had completed their dissertations in physics with acoustics topics in 1922.[2]

Trendelenburg and Meyer wrote in Waetzmann's *Technische Akustik* as well, but on entirely different subjects.[3] While Trendelenburg's *Akustik* and Waetzmann's *Technische Akustik* were very similar, there were also clear differences. The 1934 handbook included chapters on noise abatement and sound motion pictures, which did not feature as themes in the 1927 handbook, indicating the rapid development in both areas (Berger 1934; Fischer and Lichte 1934). Even more significantly, chapters on sound signaling and sound ranging in *Technische Akustik* recapitulated military research and development during World War I and reflected a newly augmented interest in the relevance of acoustics for warfare in the early 1930s (Hecht and Fischer 1934, 411–436). With the National Socialist seizure of power in 1933, the military became once again a main driving force of acoustics research in Germany. The term *Technische Akustik* derived from *Technische*

Physik (technical physics), which revolved around the relationship between technology and physics, and more generally from the discourse on Technik in the interwar scientific community in Germany.

Both the electrification and the industrialization of acoustics were directly related to the work of scientists during World War I. War research and development had introduced new technologies (especially electric amplification) and new research practices, as well as engaging acousticians with new sounds. The previous distinction of sounds into the categories of musical tones and noises had been set against the new categories of signal and noise on the battlefield. To see how these practices continued into the interwar period, we may follow the fortunes of the actors introduced in the previous chapter. Many of the authors of the German handbooks of 1927 and 1934 had worked on acoustics during World War I, and the situation was quite similar in the United States. There, the National Research Council set up a Committee on Acoustics, which addressed the transition of acoustic practices from war to peacetime in its bulletin of November 1922.[4] Several of the thirteen sections mentioned in the bulletin related directly to work carried out during the war, such as acoustics in navigation, sound propagation in the atmosphere, and George Stewart's work on conical horns. Even more than in Germany, in North America the electrical industry took a lead in postwar acoustics research and development practices. There was a large market for acoustic knowledge, but acousticians did not feel well regarded within the American Physical Society (E. Thompson 2002, 99–104). They preferred to create their own society to connect and support the community of acousticians. Scientists at Bell Laboratories in New York, the electrical industry's largest industrial research laboratory, spearheaded the establishment of the Acoustical Society of America and its *Journal of the Acoustical Society of America* in December 1928 (E. Thompson 2002, 105–107).

Although national regimes and strategies played their role, media technologies such as radio, telephony, sound motion pictures, sound recording, and sound amplification systems were spreading to all parts of the world. The dissemination and implementation of technological systems of mass media and sound amplification required acoustic knowledge to be produced locally. Premises such as picture theaters, broadcasting studios, and concert halls had to be adapted to sound recording and sound amplification across the globe. The new acoustics also required research and training

of scientists and engineers at a regional level.[5] The previous chapter showed
how scientists learned to work within the technological systems of industrial warfare during World War I; in the interwar era, they occupied a similar role in the technological systems of mass media.

Within these technological systems, the production of scientific knowledge was not a primary goal but subordinated to the overall operation of
the system. The framing of research questions and research methods were
largely determined by the technical problems that had to be solved.
Accordingly, the kind of physics practiced in these large technological
systems was largely a physics of problems rather than a physics of principles, to use a distinction made by Suman Seth in his study of Arnold
Sommerfeld (Seth 2010, 13–46). But research laboratories like Bell and Siemens also gave scientists space to follow more principle-based research
agendas. Walter Schottky, for example, one of Germany's most eminent
theoretical physicists, was engaged with fundamental questions of quantum mechanics and atomic theory. Yet he preferred the industrial research
laboratory to the university department, which came with a different set
of constraints.[6]

A number of factors created a specific German context for the emergence
and growth of Technische Akustik in the 1920s and 1930s. One of these was
the German discourse on the relationship between physics, technology
(Technik), and industry. This discourse had been integral to the establishment of the technical sciences at both the universities and the Technische
Hochschulen since the turn of the century. Other important factors
included the economic and political situation and the self-image of the
German physics community after the war. Eric Schatzberg has pointed out
that the German concept of Technik was not the equivalent of the English
term "technique" but was later integrated into the American concept of
"technology" (Schatzberg 2006). The term "Technische Physik," common
in the German community as well as in Nordic and eastern European countries, is usually translated as "technical physics," while the term "technological physics" has never entered common use. The American concept of
"engineering physics" would not be applicable, since German technical
physicists generally went to great lengths to argue that they were not engineers but physicists. The term "industrial physicists" probably comes clos-
est, since technical physicists saw themselves as practicing physics in an
industrial setting.

The arguments put forward in the discourse on technical physics deserve a closer look. Historians of science have discussed the possible influence of the economic crisis, a zeitgeist hostile to science, and political and social instability on the development of quantum mechanics—a stance generally associated with the Forman thesis.[7] These crises and instabilities affected communities of experimental and applied physicists to a similar degree as they affected theoretical physicists. Like German society at large, the physics community felt beleaguered after the German defeat. During the war, scientists had worked in military research and in the development and application of war-related technologies. Physicists saw themselves as having contributed substantially to Germany's war effort. In their view, the discipline of physics had not only advanced because of the war, but had revealed its technological and industrial potential (Hoffmann and Swinne 1994, 15). In the west, the occupation of the Rhineland and the Ruhr valley threatened German control of the coal mines that had been the life blood of many of its heavy industries. War reparations and the loss of the colonies added to the economic constraints, at least as perceived in German public debate. The only way the economy could be rebuilt, scientists argued, was on the basis of an industry founded on scientific research rather than on natural resources.[8]

German society and its scientific community felt a sense of economic, political, and moral defeat after the armistice and the Versailles Peace Treaty. In the infamous "Aufruf an die Kulturwelt," or "Manifesto of the Ninety-Three," of October 1914, German academics, among them prominent physicists including Philipp Lenard, Max Planck, and Wilhelm Conrad Röntgen, had declared their unlimited support for the German war effort (Ungern-Sternberg and Ungern-Sternberg 1996). While German scientists saw the treaty as a deep injustice against Germany, French and British scientists distanced themselves from those of their German colleagues who had signed the "Manifesto of the Ninety-Three." In the first decade after World War I, the International Research Council founded by the Allies in 1919 excluded German physicists from most international scientific collaborations, for example, the Solvay Conferences. When German scientists were invited back into international collaborations and meetings, many of them refused in protest.

Against the backdrop of these emotions, as well as the economic, political, and social realities in Germany immediately after the war, Georg

Gehlhoff founded the Deutsche Gesellschaft für technische Physik (German Society for Technical Physics) and the journal *Zeitschrift für technische Physik* in 1919. The Society for Technical Physics shared several traits with the Acoustical Society of America, founded ten years later. The society and its journal took off immediately, with a rapidly growing number of members—most of them working in industry, like Gehlhoff, who worked in the glass industry after studying physics at the University of Berlin and the Technische Hochschule Danzig (Gdańsk).[9] According to Gehlhoff, in 1919 two-thirds of Germany's physicists worked not in the public or academic sector but in industry.[10] Raising the status of the industrial physicist was therefore an important motivation for founding a separate society for technical physics. Gehlhoff and his collaborators attempted to establish a community and a discipline distinct from general physics, giving an already existing community of physicists its forum and its institutions (Hoffmann and Swinne 1994, 11).

Gehlhoff and other mentors of technical physics argued that the discipline had an identity essentially different from applied physics and from engineering. Technical physicists saw themselves first as physicists, equipped with the entire body of physical knowledge and driven by a scientific spirit that they believed was unique to the discipline of physics. At the same time, they called for the training of technical physicists to be close to the engineering disciplines, which should be subsidiary subjects in the curriculum. Graduates of technical physics should be prepared to work in industry. Technical physics should be a "technical" discipline (a technological discipline in the German sense) between science and technology, directed at applying principles and practices from physics to technological development and industrial production.[11] The founders of technical physics had a clear and normative understanding of the differences between science and engineering, and they argued for a hierarchical relationship between the two, the latter being based on and inspired by the former.[12] Unsurprisingly, there was some resistance in the engineering communities against what they perceived as physicists moving into the province of engineering (Hoffmann and Swinne 1994, 14–15).

The conflict between the community of technical physics and the communities of general physics and engineering should not be exaggerated, however. Relations between the Gesellschaft für technische Physik and the Deutsche Physikalische Gesellschaft (German Physical Society) were, by and large, smooth, with many double memberships and identities.[13] Industrial research laboratories gave physicists employment while allowing them to continue their academic activities. Communities of applied and technical physicists were also not sharply separated from communities of engineers, and many members of the Gesellschaft für technische Physik and contributors to its journal were engineers by title and affiliation. While the categorical differences between science and engineering were constantly discussed, many of the actors moved quite freely between science and engineering, as well as between industry and academic institutions. Though some actors found their credibility as physicists contested, as we will see in the case of Hermann Reiher at the Technische Hochschule Stuttgart, others, such as Heinrich Barkhausen, were accepted in both worlds.[14]

Membership of the Gesellschaft für technische Physik reached a peak in the 1920s, when it surpassed that of the Deutsche Physikalische Gesellschaft.[15] By the 1930s, the Technical Physics Society's leading role had declined, but technical physics itself became useful to the National Socialist agenda, as discussed in chapter 5. After World War II, neither the society nor its journal was revived.[16]

In the topology of both technical physics in interwar Germany more generally and technical acoustics more specifically, three important types of institutions may be identified: Technische Hochschulen (university-level technical colleges), corporate research laboratories, and public research institutes. The following sections look at the research communities and environments for technical acoustics and electroacoustics in these three types of institutions.

4.2   Acoustics at the Technische Hochschule

After World War I, large numbers of students rushed to the German universities and Technische Hochschulen. Even more than the universities, the Technische Hochschulen played a key role in the vision of rebuilding Germany through science and technology. Added to the students who had just completed secondary schooling were more than four years' worth of students who had been sent to the battlefields immediately after graduating from school. All of a sudden, the situation of universities and Technische Hochschulen changed from being depleted of students, faculty, and staff to being overrun by students, a large proportion of them war veterans.[17] Many of the students wanted to study technical subjects after having come into contact with novel technologies on the battlefield, for example, operating field telephones or wireless stations. In the subsequent years, both Technische Hochschulen and universities set up chairs and curricula for technical physics (Hoffmann and Swinne 1994, 17–18; Gehlhoff 1921). At the Technische Hochschulen, this process accompanied the elevation of the status of technical physics from a teaching subject for engineering students to a discipline in its own right, located between science and technology.

The rise of technical physics at the Technische Hochschulen was not without tensions. On the one hand, many in the established physics community were suspicious of the new career path of technical physics as an addition to that of physicists graduating from the universities. What distinguished the training of technical physicists from that of university physicists? Were technical physicists scientists or engineers? How much training did they receive in engineering subjects, and did this compromise their training in general physics? Were technical physicists qualified for university faculty positions in general physics? On the other hand, the engineering disciplines tended to perceive technical physics as a competitor, if not an actual threat, to their own field and graduates. Engineers did not necessarily think that engineering should be guided by science or that physicists were the better engineers. After all, they argued, physicists lacked practical experience and knew nothing about engineering design.

Technical acoustics was one of the promising new fields of technical physics. Before the war, only a small number of scientists had followed a research agenda in acoustics, which was seen as outdated. In the 1920s, the situation was rather different. Radio, sound amplification, and sound motion pictures created a demand for acoustics research and experts in both industry and the public sector. Technical acoustics and electroacoustics became prominent in electrical engineering as well. Whereas some engineering professions, such as civil engineering and shipbuilding, were further removed from physics, the disciplinary boundaries between technical physics and electrical engineering were not sharply drawn. Electrical engineering had been dominated by power engineering before the war, but during and after the war, high-frequency engineering, radio, telephony, and signal amplification gained prominence. This new field of electrical engineering was labeled *Schwachstromtechnik* (low-current electrical engineering) in German, in contrast to *Starkstromtechnik* (electrical power engineering).

In the course of the 1920s and 1930s, departments of physics and electrical engineering introduced research and teaching activities in technical acoustics and electroacoustics at virtually all Technische Hochschulen. The universities of Jena and Göttingen, which had taken a lead in technical physics, played only a secondary role in the agenda of technical acoustics in the interwar period. Max Wien's pioneering work in acoustics had been carried out in the prewar period, and he turned to other research fields in Jena after the war when acoustics research began to burgeon (Wagner 1937). After Hermann Theodor Simon died in 1918, his former assistant Max Reich acted as head of the Institute for Applied Electricity in Göttingen and became Simon's successor in 1920. Reich continued to work in electroacoustics as part of a larger agenda in electrical research. Under Reich, however, the institute never regained the central position it had held under Simon's prewar aegis, when it had turned out most of the German scientists working on sound location and communication during the war.[18]

Erich Waetzmann at the Technische Hochschule Breslau (Wrocław), Jonathan Zenneck at the Technische Hochschule Munich, and Ernst Lübcke at the Technische Hochschule Braunschweig were among the physicists pursuing research in technical acoustics at the Technische Hochschulen. The scientists who engaged in electroacoustics within electrical engineering included Heinrich Barkhausen at the Technische Hochschule Dresden, Hermann Backhaus at the Technische Hochschule Karlsruhe, and Richard Feldtkeller at the Technische Hochschule Stuttgart.[19] I now focus in more detail on the Technische Hochschule Dresden, where Heinrich Barkhausen held the chair in low-voltage engineering, and the Technische Hochschule Munich, which was the first engineering college to offer a degree in technical physics. These two institutions serve as examples of how scientists established research and teaching in electroacoustics and technical acoustics within a larger agenda of technical physics and electrical engineering.

4.2.1   Electroacoustics as Low-Voltage Engineering in Dresden

Barkhausen was appointed professor of electrical measurement, telegraphy, and telephony at the Technische Hochschule Dresden in 1911. He soon redefined his field of research and teaching as Schwachstromtechnik. Barkhausen's 1906 dissertation at the Institute for Applied Electricity in Göttingen laid out the framework for his overarching research program of Schwingungsforschung, or oscillation research, which he followed throughout his career in Dresden. Barkhausen regarded the generation and continuation of oscillations as the foundational principle that brought together the fields of electromagnetism, mechanics, and acoustics (Barkhausen 1907; Lunze 1981, 15).

In his 1911 inaugural lecture at Dresden, Barkhausen made it clear that acoustics would become a central and integral part of his plans for electrical engineering. He pointed out that the development of the telephone and microphone had not hitherto been based on scientific research, in contrast to wireless telegraphy, which he saw as developed out of scientific theory and practice. Barkhausen downgraded the prewar carbon microphone to basically nothing but a primitive loose contact. The phonograph was the one sound technology that fared worse in Barkhausen's lecture than the microphone and telephone. He declared that only the human ear in combination with human intellect could construct words and sentences out of the horribly distorted sounds coming from the machine (Barkhausen 1911, 516–517). This contrasted starkly with the view of Carl Stumpf, Otto Abraham, and Erich Moritz von Hornbostel, who had credited the phonograph with superior scientific objectivity compared with the human listener. According to Stumpf, the phonograph could reproduce exactly both rhythmic and tonal properties of music and preserve them unchanged.[20] For Barkhausen, however, the tonal properties of the music were not reproduced exactly by the phonograph, but only in the human ear and mind.

If Barkhausen's dissertation was the blueprint for his scientific program at Dresden, his research and development at the Torpedo-Versuchs-Kommando between 1915 and 1918 provided him with practical experience. Investigating sound propagation in the sea had exposed Barkhausen to the problems of electroacoustic transducers and sound measurement. His investigations of underwater sound, along with electromagnetic telegraphy, were at the core of the analog understanding of electromagnetic and sound wave propagation, as discussed in chapter 3. In 1917, Barkhausen started to investigate systematically the properties of amplifier tubes and circuits. He continued his research on tubes and amplifiers into peacetime, resulting in the 1920s in a series of textbooks that were reprinted in many revised editions and that made Barkhausen a German authority on the subject of radio tubes.[21]

Returning to Dresden after the war, Barkhausen integrated acoustics research into his broader agenda of electrical engineering from the mid-1920s onward. His plans for electroacoustics differed decisively from acoustics research practices before the war (Reichardt 1981).[22] When Barkhausen was appointed at the Technische Hochschule Dresden in 1911, Richard Heger, honorary professor of mathematics, had been lecturing in architectural acoustics there. Heger measured reverberation time and experimented with various sound-absorbing materials while providing consultancy services in architectural acoustics. Having taken this initiative, Heger was appointed head of the Sammlung und Arbeitsstelle Raumakustik (Collection and Office of Architectural Acoustics) at the Technische Hochschule Dresden in 1912.[23]

Although Heger based his approach to architectural acoustics on Wallace Sabine's groundbreaking work in the field, he was doubtful how far Sabine's results could be generalized. Sabine had used the tone of an organ pipe in his reverberation measurements, which produced one single frequency of 512 cps; Heger used a popgun with a wooden plug, which produced a broader acoustic spectrum. Heger acknowledged that the human ear was the most sensitive sound detector available to him; however, he saw the human observer as highly problematic, since sound sensitivity varied from day to day even for the same observer. Heger tried to record reverberation mechanically on a gramophone, but he could not find a membrane sensitive enough for his purposes. According to his 1911 article, Heger used a telegraph for timekeeping but no other electric instruments, such as microphones or galvanometers, as George W. Pierce at Harvard had done in 1908 (Heger 1911).[24] Heger retired in 1917 and died in 1919, when Barkhausen had just come back from the war. There is no indication that Barkhausen continued the work of Heger, and the Collection of Architectural Acoustics was left "orphaned" in the early 1920s.[25]

Barkhausen established a rather successful school of low-voltage engineering that had international impact. Similar to other academic schools of the time, such as the interwar schools of quantum mechanics, Barkhausen created an intimate social atmosphere, going climbing and skiing with his students, assistants, and colleagues (Lunze 1981, 31). Most noteworthy is a group of Japanese students who made the Barkhausen school the foundation stone of low-voltage engineering in Japan. Among these were Hidetsugu Yagi, who visited Barkhausen before World War I and became an entrepreneur in Japanese communication engineering; Nobushige Kato, later professor at the Imperial University of Kyoto; Yoji Ito, who translated Barkhausen's book on the amplifier tube into Japanese; and Tsonge Shih, who completed his dissertation on radio tube noise in 1937 (Lunze 1981, 56, 121, 126, 127; Wein 2011, 91–92). Norwegian students made up another large international group at Barkhausen's institute. Barkhausen visited the United States in 1929 and the Soviet Union in 1930, and he went on a lecture tour to Japan in 1938.

Barkhausen's contributions to acoustics in the 1920s included the introduction of the phon as a measurement unit for the subjective loudness of a sound, and the development of a subjective sound meter.[26] With his students G. Lewicki and Horst Tischner, he investigated the relationship between the physical intensity of complex sounds and their perceived subjective loudness.[27] In 1902, Max Wien had determined that the sensitivity of human hearing for a sinusoidal tone was greatest around 1,000 cps, and that the perceived loudness declined on a logarithmic scale with smaller frequencies (Wien 1903). For complex sound, Barkhausen, Lewicki, and Tischner found that its loudness equaled the loudest harmonics perceived, and that no other harmonics contributed.

To measure the loudness of complex sounds, Barkhausen designed an instrument that produced a buzzer sound, which was regulated with a potentiometer. The observer received the buzzer sound through a telephone receiver on one ear and compared it to the sound perceived on the other.[28] The loudness was measured by regulating the loudness of the buzzer sound until the loudness perceived on both ears was the same. For comparison, Barkhausen proposed the linear Wien scale (named after Max Wien), representing the physical sound intensity, and the logarithmic phon scale, representing the perceived loudness (Barkhausen 1926; Reichardt 1981, 81–82). Siemens and Halske started to produce the Barkhausen noise meter—Geräuschmesser für die Praxis (phonometer) nach Barkhausen—as early as 1926. Barkhausen's noise meter became a standard instrument for measuring sounds like automobile noise and in Heinrich Fassbender and Kurt Krüger's measurement of noise inside airplanes at the Deutsche Versuchsanstalt für Luftfahrt (German Testing

Laboratory for Aviation) in Berlin-Adlershof in 1927 (Meyer 1934, 138–140; Fassbender and Krüger 1927).

Beyond his own research, Barkhausen supervised many diploma and doctoral students on acoustic subjects. Martin Kluge estimated in 1941 that of around three hundred diploma theses in electrical engineering at Barkhausen's institute, more than forty were in the field of acoustics (Kluge 1941, 315). At least thirteen of the fifty-seven doctoral dissertations supervised by Barkhausen between 1925 and 1939 belonged to acoustics and its subfields (Lunze 1981, 25–27). Barkhausen's department was electrical engineering, and it is no surprise that his and his students' approach to acoustics generally was through electroacoustics. Many of the studies dealt with electroacoustic devices.[29]

For Barkhausen and his students, the meaning of electroacoustics extended beyond electroacoustic technologies to comprise an electroanalog understanding of even those acoustic oscillations not produced or measured electromechanically. Barkhausen's overarching research concept of analog oscillation research favored the approach of translating acoustic oscillations into electrical oscillations, then studying them with the help of circuit analysis and other methods from electrical engineering. Horst Tischner, Joachim Tröger, and Martin Kluge used electroanalog methods and analog circuit diagrams for nonelectrical systems. In his dissertation, Tröger worked on the mechanical impedance of the human eardrum, using analogies from electric circuit theory (Reichardt 1981, 81; Tröger 1930). Tröger and Kluge worked on reducing the noise produced by combustion engines by improving exhaust silencers; Barkhausen and his students had been studying the noise of combustion engines since 1925. As Kluge explained, an exhaust silencer acted basically as a filter, filtering out the frequencies that bothered the human ear. For his design of an automobile exhaust silencer as an acoustic filter, Kluge could draw on existing electric filter theory, and translated his acoustic system of the sound silencer into an equivalent electric circuit diagram (figure 4.1).[30]

Even though Barkhausen taught in an engineering department, it would be wrong to think of him as an engineer alone. He and his students published in a broad range of journals, from electrical engineering to general physics; made recognized original contributions to science; and presented their work at meetings of the Deutsche Physikalische Gesellschaft. This made Barkhausen, and to some extent also his students, visible members of

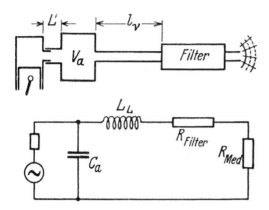

Figure 4.1
Design and electric equivalent circuit diagram of an automobile exhaust silencer with accumulation chamber by Martin Kluge (Trendelenburg 1934, 169).

the German physics community.[31] "Dual citizenship" in both the physics and electrical engineering communities in Germany seems to have been quite typical for acousticians of the interwar period, and to a great extent, Barkhausen's academic profile was shared by Jonathan Zenneck, a prominent representative of the community of technical physicists in Germany and professor of physics at the Technische Hochschule Munich.

4.2.2 Technical Acoustics as Technische Physik in Munich

While Felix Klein was establishing technical sciences at the University of Göttingen, Walther von Dyck and Carl von Linde worked along similar lines at the Technische Hochschule Munich. Dyck was a mathematician and former student of Klein's, while Linde was a mechanical engineer and industrial entrepreneur in refrigeration and the liquefaction of gases. On the initiative of Dyck and Linde, in 1902 Munich became the first Technische Hochschule in Germany to set up a laboratory for technical physics. It was the only one to award doctoral degrees in technical physics before World War I. From 1906, it awarded diplomas in technical physics as well.

Dyck and Linde managed to get the physicist Oscar Knoblauch appointed as professor extraordinarius in technical physics, in the face of resistance from an engineering faction that wanted to appoint a practitioner and not a scientist (Manegold 1970, 281–282; Hashagen 2003, 332–334). Knoblauch chose technical thermodynamics as the main research field of his laboratory. Most of Knoblauch's assistants had a first degree in engineering (Knoblauch 1941–1942, 9). In the years before World War I, there were few students of technical physics in Munich, since not many physicists worked in industry. Among the technical physicists was Hilde Mollier, who worked on the steam pressure of ammonia solutions as Knoblauch's assistant between 1905 and 1909. She was one of the first women to study at the Technische Hochschule Munich, and was Dyck's cousin. Hilde Mollier married Heinrich Barkhausen in 1909, which ended her own scientific career. Like many wives of professors, she supported Barkhausen in his career and assisted him in Dresden with setting up student laboratories and with his paperwork.[32]

Investigations of heat conduction and insulation in buildings drew Knoblauch's doctoral candidates to sound propagation and sound insulation. Both heat and sound were important variables in the physics of buildings, and their propagation and insulation could be dealt with in analogy. The first of these candidates was Richard Berger, who wrote his dissertation on sound transmission and insulation in buildings between 1907 and 1911 (Berger 1911). Berger worked on sound ranging of artillery during World War I and started to teach at the Gauß Engineering College in Berlin in 1927 (Knoblauch 1941–1942, 24; Berger 1918). In 1928, he initiated the journal *Die Schalltechnik*. Knoblauch continued Berger's acoustic investigations by appointing Rudolf Ottenstein to work on acoustics between 1911 and 1914, Ernst Schmidt between 1914 and 1923, and Hermann Reiher between 1922 and 1929. Whereas Berger worked only on sound, Schmidt and Reiher worked on both sound and heat insulation.

Knoblauch represented technical physics but did not hold the more prestigious main chair of physics at the Technische Hochschule Munich. That was held by the professor of physics who taught general physics to the engineering students. In 1913, Jonathan Zenneck was appointed to succeed Hermann Ebert as the main chair of physics and head of the physical cabinet. Like Knoblauch, Zenneck picked up technical acoustics in the course of his wider research agenda in Munich. But whereas for Knoblauch acoustics became part of building physics, for Zenneck it became part of an agenda in electroacoustics and high-frequency physics. As with Max Wien and Heinrich Barkhausen, Zenneck's general research approach can be characterized as oscillation research, bringing together acoustics and electric oscillations. But for Zenneck, oscillation did not become the fundamental and overarching principle that it was for Barkhausen and others. Zenneck had worked as Ferdinand Braun's assistant in Strasbourg between 1895 and 1905. He published papers on the acoustic vibration of plates (Zenneck 1898, 1899a) and other mechanical oscillations, but was generally known for his pioneering research on electromagnetic waves and wireless telegraphy. He had carried out experiments with the Braun wireless transmitter on ships on the North Sea in 1900. In the preface of his textbook *Elektromagnetische Schwingungen und Drahtlose Telegraphie* (Electromagnetic oscillations and wireless telegraphy) of 1905, Zenneck argued that he had limited the use of mechanical analogies because he wanted the reader to understand electromagnetic phenomena in their own right (Zenneck 1905, vii). He could not resist, however, pointing out the "complete analogies" between acoustic and electric oscillations in the case of resonance (328, 556). Zenneck pioneered the use of Braun's cathode ray tube as an alternative to the rotating mirror oscillograph in visualizing and measuring high-frequency oscillations.[33] Many of his publications showed photos of oscillations visualized by a Braun tube.

Zenneck was appointed professor extraordinarius at the Technische Hochschule Danzig in 1905, and full professor in Braunschweig in 1906. In 1909 he joined BASF, the Badische Anilin- und Soda-Fabrik, where he worked on processes oxidizing nitrogen from the air by an electric arc. In 1911, Zenneck moved back to the Hochschule in Danzig, this time as the successor of Max Wien (who had left for Jena), before finally moving to Munich in 1913. With his assistant Hans Rukop, Zenneck investigated the oscillations of the electric arc in 1913 and 1914.[34] Like many German physicists, Zenneck spent the first months of World War I on the battlefield. As a reserve officer, he commanded a company of naval infantry in Belgium until the German authorities decided that his qualifications would be more useful in a different mission. Zenneck and Ferdinand Braun traveled to the United States in December 1914 as expert witnesses for Telefunken in a patent lawsuit brought by the Marconi Company. Before leaving, Zenneck communicated his proposal for sound ranging of artillery to his assistant Ernst von Angerer.[35] The Marconi lawsuit threatened to close down the Telefunken station at Sayville on Long Island, which was crucial to Germany's global wireless network and therefore also to German warfare. Suspected of being a spy, Zenneck was detained on Ellis Island when the United States entered the war in 1917. He was released only in 1919, then returned to his professorship in Munich (Zenneck 1960, 301–395).

After the war, student numbers in technical physics in Munich shot up (Knoblauch 1941–1942, 6; Zenneck 1960, 396). New degree regulations for the diploma in technical physics were introduced in March 1924, splitting the study of technical physics into two separate tracks: Track A emphasized engineering, Track B, physics. Track A included machine design and technical drawing, while Track B included theoretical physics in the compulsory curriculum. As Zenneck put it, Track A was tailored to Knoblauch's students in the Laboratory for Technical Physics, while Track B was a general physics curriculum similar to that offered by the universities and favored by Zenneck himself.[36] In 1925, 67 students were registered in technical physics at the Technische Hochschule Munich. Student numbers climbed to 140 by 1930, then declined again to 53 in 1937.[37] By 1935, 100 students had graduated with a diploma in technical physics (38 from Track A and 62 from Track B), indicating that many registered diploma students did not graduate. From 1920 to 1933, 20 mechanical and electrical engineers completed doctoral dissertations in technical physics.[38]

Because of his stay and detention in the United States, Zenneck could not capitalize on war research and development as Barkhausen and others did. For Zenneck, electroacoustics was part of a larger agenda of high-frequency physics, or radio physics, just as for Barkhausen, it was part of a larger agenda of electrical engineering. With Eberhard Mauz of the Physikalischer Verein (Physical Society) in Frankfurt am Main, Zenneck edited the *Zeitschrift für Hochfrequenztechnik* (Journal for high frequency technology). In 1932, Mauz and Zenneck acknowledged the growing importance of electroacoustics by adding the term to the journal's title, now *Zeitschrift für Hochfrequenztechnik und Elektroakustik*. When Kurt Jacoby, the publisher, proposed placing contributions to electroacoustics in separate supplements to the journal, Mauz and Zenneck dismissed the idea. In many cases, they argued, electroacoustics was inseparable from high-frequency physics, especially when publications discussed amplifier valves and circuits.[39]

Alongside the broader context of radio physics, Zenneck developed a specific interest in architectural acoustics (Schmucker 2000, 419–424). In a résumé of 1952, Zenneck placed architectural acoustics second among his fields of work, after electromagnetic radiation and before his ionospheric research.[40] Zenneck's engagement in architectural acoustics can be traced back to at least 1927 and included supervision of a considerable number of diploma and doctoral students.[41] Although Hermann Reiher at the Laboratory for Technical Physics published on architectural acoustics around the same time, there are no signs of collaboration between Zenneck and Reiher or any other researchers at Knoblauch's laboratory.[42] Zenneck and his students investigated the acoustics in a variety of buildings, starting with the large physics lecture hall and other premises at the Technische Hochschule in Munich. The investigations were funded by the Notgemeinschaft der Deutschen Wissenschaft (Emergency Association of German Science) and the Helmholtz Society. The acoustic investigations of his doctoral students Walter Schindelin and Ernst Scharstein, published in *Annalen der Physik* in 1929, included the Munich broadcasting studio and the assembly hall of the University of Freiburg (Scharstein 1929; Schindelin 1929; Scharstein and Schindelin 1929). The investigations of Wolfgang Linck and Walter Kuntze included the Hall of Fame at the Deutsches Museum and the reservoir tank of the Walchensee hydroelectric power station, which was chosen because of its extraordinarily long reverberation time (Linck 1930; Kuntze 1930).

Zenneck and Schindelin argued that the problems of architectural acoustics could only be tackled appropriately using electroacoustic measurement technology, which enabled acousticians to measure sound with objective methods. The ear was the ultimate judge of architectural acoustics, but it could not adequately detect the causes of poor acoustics (Schindelin 1929, 129). Zenneck's students produced shock sounds by firing pistol shots and tones of different frequencies with a tone generator and a loudspeaker. They recorded the resulting reverberation via microphone on a Siemens loop oscillograph.[43] While most acousticians published diagrams of reverberation time, Zenneck and his students published and discussed only their oscillograms; mathematical analysis was largely absent.[44] Instead, Zenneck and his students focused on the directivity of sound rays and the directional characteristics of listening. The location of speakers, musicians, and listeners was at least as important for them as the shape of reflecting surfaces and the location of sound-absorbing materials.[45]

Zenneck's activities in improving the acoustics of the Munich Prinzregententheater brought him into contact with the Hamburg Musikhalle, a neobaroque concert hall (Crone, Seiberth, and Zenneck 1934). Zenneck sent his assistants Willy Crone and Hans Seiberth to Hamburg to carry out the measurements and recommend modifications in 1934. Zenneck and his assistants recommended installing curtains and other sound-absorbing fabric at the places they had identified as sources of harmful reflections. Musicians and music critics, however, reacted rather negatively to the acoustic modifications of the Musikhalle. Conductor Hans Hoffmann found the new sound dull and hollow, a mushy sound impression that lacked all brilliance. The composer Hermann Erdlen and musical director Richard Richter made similar judgments. They all recommended removing the curtains that had been put up following the instructions by Zenneck and his assistants and proposed to improve the acoustics by replacing the glass ceiling with a wooden ceiling.[46] The expertise of the scientists clearly conflicted with that of the musicians, who did not seem to require an oscillogram for their evaluation. Zenneck responded to the criticism by admitting that the acoustic deadening of the hall might have gone too far and that some of the curtains should probably be removed. As the scientific authority on acoustics, however, he recommended approaching "the musicians' criticism with some suspicion," and added that the acoustics of a concert hall would always be a compromise.[47]

Zenneck continued to consult on architectural acoustics. Most noteworthy among his projects were the Musensaal of the Mannheimer Rosengarten concert hall in 1934, the Theater Hall at Adolf Hitler's mountain residence on the Obersalzberg in 1937, and Bavaria Filmkunst's synchronization studio for sound motion pictures in Munich in 1939.[48] Zenneck was probably motivated to enter architectural acoustics by his interest in the acoustics of speech and music. He had spent much effort in planning the Technische Hochschule Munich's new physics auditorium to accommodate the increased numbers of students. Speech intelligibility was an important component of the lecture hall's success.

Architectural acoustics also fitted well with the dynamics of research and training students. Though Zenneck did not collaborate with other scientists in architectural acoustics, not even within his own institution, he organized the work of his students in a highly collaborative way. All dissertations built on one another and used one another's resources. Similarly, Knoblauch ensured the continuity of the acoustics agenda in the Laboratory for Technical Physics. But beyond personal interest, it made sense for both Knoblauch and Zenneck to accommodate technical acoustics within research and teaching in technical physics. If the curiosity of scientists and students was important for the rise of technical acoustics at the Technische Hochschulen, even more decisive was the demand for acoustical expertise and experts created by the rise of radio broadcasting, sound motion pictures, sound amplification, and new practices in the construction of buildings.[49] Many, if not most, of Zenneck's students eventually worked in the electrical industry, and Zenneck corresponded with companies to help place them.[50]

While architectural acoustics was an important scientific agenda for Zenneck, it would never be his only field of research. Zenneck himself argued that the background of physicists who had spent most of their career in industrial laboratories was too narrow for professorships at universities and Technische Hochschulen. He made this explicit in his evaluations of Ferdinand Trendelenburg at Siemens Laboratories and his own former assistant Hans Rukop, who was now head of research at Telefunken, as professorial candidates. Zenneck evaluated Erwin Meyer, who had worked at the Telegraphentechnisches Reichsamt and the Heinrich Hertz Institute, in a similar way.[51] Trendelenburg and Meyer were arguably the leading German experts in the new acoustics and had a strong research record, but this did not necessarily make them good professorial material in Zenneck's eyes.

While physicists in industry and at other research institutions could be highly specialized, it was still expected that a physics professor at a university or a Technische Hochschule would be more of a generalist, follow several research agendas, and represent the entire discipline. The same applied for a professor of electrical engineering like Barkhausen, for whom electroacoustics was only one activity within his larger research and teaching in low-voltage electrical engineering. The cases of Zenneck and Barkhausen recall Michael Aaron Dennis's characterization of Charles Stark Draper and his establishment of the MIT Instrument Laboratory in Cambridge, Massachusetts, during the 1930s: "In the laboratory's crowded space, Draper developed a distinctive pedagogical style, while producing both students and novel measurement instruments" (Dennis 1991, 24). Draper's aeronautical instruments had been developed from educational laboratory devices into instruments that were used in airplanes and ultimately produced commercially as aviation instruments. Dennis describes teaching as Draper's central identity and approach, even in his work as a researcher and consultant. Just as for Draper, for Zenneck and Barkhausen, the students remained their most important resource in a larger framework of teaching, research, and consulting.

4.3   Sound Industry: Acoustics Research in Corporate Laboratories

Research in the corporate laboratory differed significantly from research and teaching at the Technische Hochschule. Acoustics research in the electrical industry was a new phenomenon, emerging largely during World War I. Before the war, research in the electrical industry concentrated on heavy-current engineering, electric lighting, and wireless telegraphy. The development of the phonograph and the telephone had not been based on scientific research, as Barkhausen pointed out in 1911. Until 1914, it was not scientists but inventors and industrial entrepreneurs like Thomas Edison, Alexander Graham Bell, and Emil Berliner who dominated the development of electroacoustic technologies. Some research had been carried out on long-distance telephony, but little on the acoustic properties of the telephone.[52]

During World War I, underwater acoustics for submarine warfare and the development of microphones, amplifier tubes, and amplifiers brought scientists working in the electrical industry to acoustics. Hans Riegger and Hans Gerdien worked on underwater acoustics and the development of microphones and sound transmitters for Siemens, while Walter Hahnemann and Heinrich Hecht carried out similar research for the Signal Gesellschaft. During and after the war, research and development in the electrical industry experienced a transition from analysis and testing to systematic research in larger central research laboratories. While industry had already employed scientists before the war, for example, Barkhausen and Zenneck, their research efforts now became more structured and long ranging. Research was not necessarily directed toward a certain product: scientists and industrialists argued that corporate laboratories also needed to research the scientific foundations of the technologies and methods on which the company's products were based. Siemens opened its research laboratories in the Siemensstadt district of Berlin in 1920 (Trendelenburg 1975, 46). The Dutch Philips corporation, one of Siemens's largest competitors in Europe, set up a new building for its Natuurkundig Laboratorium (Natural Science

Laboratory, NatLab) in 1922. The Allgemeine Electricitäts-Gesellschaft (AEG) was rather late in comparison to Siemens, opening its research laboratory only in 1928. Under the leadership of Carl Ramsauer, however, the AEG Research Institute soon became one of the most prestigious and powerful institutions in the German industrial research landscape.

The relationship between Technische Hochschulen and industry was necessarily a symbiotic one. Electrical and other industries required highly qualified engineers and scientists. At the same time, producing these highly qualified engineers and scientists for industry was the main reason the Technische Hochschulen existed in the first place. While physicists at universities usually spent their whole academic career inside the university system, professors of technical physics and engineering at the Hochschulen preferably had some years of industry experience.[53] Physicists at universities were often uncomfortable with the patents system or receiving financial profit from industry, finding it inconsistent with the ethos of science.[54] Scientists at a Technische Hochschule, such as Barkhausen and Zenneck, filed patents and maintained close contacts with industry, which helped them to place their candidates after graduation.

For physicists, employment prospects and career paths looked distinctly different before and after World War I. Whereas industrial physicists had been the exception before the war, after the war, they became the norm. Physicists who graduated from universities around 1900 mainly went into academia or school teaching (Kragh 1999, 13). Helga Schultrich has shown that by 1930, although technical physicists worked in the public sector as well, the majority of physicists worked in industry (Schultrich 1982, 212–218). Many physicists stopped active research when they entered industry and went into management and administration, but some enjoyed quite successful research careers in industry. In both the United States and Germany, a new generation of acousticians was defined by their activities in the corporate world rather than by their university affiliation. They did not see their involvement with industry as a threat to their academic standing but saw it as a beneficial or at least necessary part of their scientific activities.[55]

A variety of reasons could make a corporate research laboratory an attractive employer, sometimes even more attractive than prestigious faculty positions. Walter Schottky, for example, moved back to the university after having worked for Siemens during World War I. He kept up his ties with

Siemens and returned fully after 1927 (Serchinger 2008, 159–275). Industry could pay higher salaries, did not tie down scientists with teaching obligations, and offered well-equipped research facilities. The problems to be solved and the industrial work environment could be stimulating and creative. Many of the research technologies of the interwar period, including particle accelerators, electron microscopes, and ultracentrifuges, were developed in industrial research laboratories or in close proximity to industry (Joerges and Shinn 2000).

To be sure, academic freedom had to be negotiated with the company in order to protect the company's interest. Patent-sensitive information could not be published. Large companies produced their own scientific publications, which boosted their scientific credentials but also enabled them to control the spread of potentially sensitive information. Julius Springer, one of the leading publishing houses for science, published the *Wissenschaftliche Veröffentlichungen aus dem Siemens-Konzern* (Scientific publications of the Siemens Company) as well as the *Jahrbuch des Forschungs-Instituts der Allgemeinen Electricitäts-Gesellschaft* (Yearbook of the AEG Research Institute). Not surprisingly, scientists at corporate laboratories accounted for a large portion of the acoustical research carried out between the wars. Corporate researchers in the United States and in Germany, such as Ferdinand Trendelenburg, Harvey Fletcher, and Hugo Lichte, produced not only journal articles, but also leading textbooks and handbooks of the field.[56]

The United States was acknowledged as the global leader in the field of acoustics research in the interwar period.[57] In North America, the laboratories of the electrical industry were more dominant than in Germany, where the central role was played by public research institutions, especially the Heinrich Hertz Institute.[58] Bell Laboratories dominated research in the United States through the strong monopoly position of American Telephone and Telegraph (AT&T) in the field of telephony. AT&T and its subsidiaries operated more telephones and more kilometers of telephone lines than the rest of the world combined, with Germany taking up second place. Americans also owned more radio transmitters and receivers than the rest of the world put together.[59] This strong electrical communication industry reflected research and development efforts. Comparing the workforce of the electrical industry's research laboratories and state laboratories in the United States and Germany in 1925, Friedrich Wilhelm Hagemeyer found that the research laboratories of Siemens and Halske were comparable in size to the General Electric research laboratory, but Bell Laboratories dwarfed both. While the *Zentrallaboratorium* (central laboratory) of Siemens and Halske had a staff of 300, 140 of them scientists and engineers, Bell had a staff of 3,500, of whom 1,400 were scientists and engineers.[60] At least 80 percent of the members of the Acoustical Society of America were affiliated not with academic institutions but with corporations offering various kinds of acoustic products and services. These included businesses in the electrical industry, construction industry, and sound motion picture industry (E. Thompson 2002, 105, 356n143).

The research laboratories of the interwar electrical industry all had their acoustics specialists and research groups. The reason for this investment in acoustics research was plain, since electroacoustic media technologies had started to become a large market. Three large technological systems drove the economy for electroacoustic technologies in the interwar period: telephony, radio broadcasting, and sound motion pictures. All three systems led to the production and consumption of acoustic knowledge on various levels. The role of the electrical industry within the large technological system was mainly to develop, produce, and supply electroacoustic devices and other hardware. But to do this successfully, and to survive economically in an aggressive and competitive market, the industrial actors had to develop an understanding of the large technological system and its entire dynamics.

In Europe, the dynamics of telephony and radio broadcasting was rather different from those in the United States. American telephony was completely dominated by the monopoly of AT&T, a government-controlled private corporation, whereas in Germany and in most other European countries, the telephone monopoly was held by the state.[61] Siemens and Halske was the main producer of telephones and other telephony hardware for the Reichspost (German Post Office), which administered and maintained the state monopoly not only on postal services but also on telephony and telegraphy. The situation was similar for radio broadcasting. Radio was also an area of private enterprise in the United States, the electrical and music recording industries being important players. In Europe, radio was largely state owned and state controlled. In Germany, the Reichs-Rundfunk-Gesellschaft (German Broadcasting Corporation) formed a national umbrella for the regional broadcasting companies. The Reichspost also controlled the Reichs-Rundfunk-Gesell-schaft. In 1920, the Reichspost founded the Telegraphentechnisches Reichsamt, which was responsible for engineering, technical operation, and organizing research and development in the fields of telephony, telegraphy, and radio broadcasting.[62]

The dynamics of the growth and dissemination of sound motion pictures, which was not regulated by a state monopoly, was very different. Private production companies divided up the international market on both sides of the Atlantic, while the electrical industry developed and produced sound recording, reproduction, and amplification equipment. In Berlin, engineers Hans Vogt and Joseph Massolle and physicist Josef Engl formed the inventor consortium Tri-Ergon to establish sound motion pictures with a sound-on-film system in the early 1920s. American inventor Lee De Forest came to Berlin around the same time to work on his Phonofilm sound motion picture system. Neither Tri-Ergon nor De Forest managed a breakthrough with their system. In the late 1920s, Warner Brothers introduced feature-length sound motion pictures on the Vita-phone system developed by Western Electric. Tri-Ergon and Phonofilm failed not because of the technologies themselves, but because neither attracted the kind of capital nor created the alliances in the film production and electrical industry that would have been necessary to reorient the existing system of motion pictures from silent to sound (Jossé 1984; Mühl-Benninghaus 1999).

The large sound-technology systems of the interwar years—telephony, radio broadcasting, and sound motion pictures—were mostly separate from each other, not unlike the case of military surveying, submarine warfare, and aerial warfare as large technological systems of warfare. However, electroacoustic sound technologies and acoustic knowledge and practices transcended the boundaries of these systems and established links between them. All three systems required technologies of sound recording, sound transmission, sound reproduction, and sound amplification. Yet, sound recording and sound amplification were not always part of a large technological system. They could also have a life of their own. Amplification systems for lecture halls, churches, theaters, and concert halls could, but did not necessarily have to, be used in radio broadcasting or sound motion pictures. Both large technological systems and smaller independent systems all required microphones, amplifiers, and loudspeakers, which were identical or shared many specifications and were generally produced by the same electrical companies. Sound recording and amplification units were assembled on a modular basis and often included standardized instrument racks.[63]

The organization and dynamics of research and development in the electrical industry was distinct from research and teaching at the Technische Hochschulen and universities. Research in industry was oriented not on disciplinary boundaries and graduate students, but on products and technological systems, small and large. Corporate scientists needed to understand and develop the scientific principles on which their corporation's present and future products were and would be based. Acousticians at corporate laboratories were concerned with the principles, characteristics, and design of microphones and loudspeakers, electric signal amplification, and amplifier tubes and circuits. Complex electroacoustic systems required sound engineers who could control sound input and output and mix sounds from different sources. These engineers dealt with frequency response, distortion, feedback, and noise that had their origin in the electric components and circuits. Acousticians at corporate laboratories studied speech intelligibility, human hearing, and music perception in order to optimize their systems technologically, but also economically. How much signal distortion was acceptable on telephone lines? To what extent could music signals be compressed in sound storage and reproduction before the human ear could hear the difference between original and reproduction? Last but not least, scientists and engineers in the electrical industry dealt with sound propagation in the atmosphere, architectural acoustics, and noise abatement, all of which were critical to the performance of their sound systems.

The big players in electroacoustic research in Germany during the interwar period were Siemens and Halske and AEG. Among the smaller companies were C. Lorenz AG and Körting Radio Werke, which both produced radio transmitters and loudspeakers. While the radio business thrived in the 1920s, underwater acoustics in Germany stagnated because of the U-boat ban and the general restrictions on the size of the navy. The Boston Submarine Signal Company gave Atlas Werke financial support, while Neufeldt and Kuhnke could not hold on to Signal Gesellschaft, which was sold to Atlas and finally liquidated in 1926 (Rössler 2006, 25). In the same year, Heinrich Hecht and others from Signal Gesellschaft founded a new company in Kiel, Electroacoustic GmbH (Elac), which was in certain ways a successor to Signal Gesellschaft. Walter Hahnemann had moved on to C. Lorenz AG.

The German electrical industry's largest international competitors were the American companies, especially Western Electric (a subsidiary of AT&T), General Electric (GE), and the Radio Corporation of America (RCA), which was created on the initiative of the United States government and owned by GE and Westinghouse. The American companies participated in the German market as shareholders, through patent licensing and agreements, and through cooperations. One of the strongest European rivals was Philips. The acoustics group of the Philips NatLab was closely linked to its radio research and development. In 1931, the laboratory's acoustics group included three scientists, six engineers, sixteen assistants, and a staff of twenty-one in the workshop. Among them were Gilles Holst, who was the head of the NatLab, and Maximilian J. O. Strutt (Boersma 2002, 59, 86), who wrote the chapter on architectural acoustics in the *Technische Akustik* handbook of 1934—the only contributor to the volume not based in Germany (Strutt 1934).

Though Siemens and AEG were competitors on the market for heavy-current engineering and other fields, their corporate leadership cofounded several subsidiary companies, among them Telefunken and Klangfilm. Telefunken was founded in 1903, partly on the initiative of Kaiser Wilhelm II and the German military authorities, to avert competition on the German market for wireless communication. In the agreement between AEG and Siemens, Telefunken took over the development and production of wireless communication systems. With the advent of public radio broadcasting, the market changed considerably in the 1920s, when AEG, Siemens, and Telefunken all produced electroacoustic devices such as radios, amplifiers, and loudspeakers. Telefunken managed international patent rights for the consortium, such as the Rice and Kellogg electrodynamic loudspeaker patents.[64] Siemens and Halske and AEG founded Klangfilm GmbH in 1928 to consolidate their efforts to develop a German system for sound motion pictures that could compete with the U.S. electrical and sound motion picture industries and prevent them from taking over the market completely. In 1931, Siemens, AEG, and Klangfilm transferred most of their research and development in electroacoustics to Telefunken.[65] Both AEG and Siemens, however, continued research and development in electroacoustics in the

1930s, for example, when AEG developed electromagnetic tape recording. The coownership of Telefunken by Siemens and AEG created considerable tension, the result of different company structures and fierce competition between the two. After several years of negotiations, Telefunken was taken over fully by AEG in 1941.[66]

4.3.1 Electric Noise: Amplifiers and Loudspeaker Design at Siemens Laboratories

The Siemens industrial concern, consisting of Siemens and Halske AG and Siemens Schuckertwerke, operated the largest corporate electrical research laboratories in Germany. Siemens developed and supplied most of the hardware for the telegraph and telephone network operated by the Reichspost state monopoly. Telephony, amplifiers, and amplifier tubes, along with microphone and loudspeaker design, dominated acoustics research at the Siemens laboratories in the interwar period. Its trajectory was in several ways shaped by the company's military engagement during World War I.

Scientific research had a long history at Siemens. Werner Siemens, who established the company with Johann Georg Halske in 1847, carried out his own physical research and was one of the main actors behind the foundation of the Physikalisch-Technische Reichsanstalt. His successor Wilhelm von Siemens intensified research efforts and employed a number of physicists, among them Hans Gerdien, a student of Woldemar Voigt in Göttingen, and some of Hermann Theodor Simon's students from the Göttingen Institute for Applied Electricity, among them Heinrich Barkhausen, Wilhelm Rihl, and Karl Boedeker. Wilhelm von Siemens started to plan a research laboratory for basic research in physics and chemistry in 1912, and he commissioned Gerdien to carry out the execution. World War I delayed the project, but the Physical-Chemical Research Laboratory was finally founded in 1919 and moved into a large new building in Siemensstadt, Berlin, in 1920. It changed its name to Forschungslaboratorium der Siemens & Halske AG und der Siemens-Schuckertwerke GmbH in 1924. Most of the research at Siemens in the interwar period was related to low-current engineering. Heavy-current engineering, which made up a large part of Siemens's actual business, was not strongly represented (Trendelenburg 1975, 46). In fact, the research laboratory never had the monopoly on research and development at Siemens. The central laboratory and the physical laboratory of the Siemens measuring technology department Wernerwerk M both engaged in their own independent research into loudspeakers and microphones, rivaling the scientists at the research laboratory. In 1920, Siemens also started to publish its own research journal, *Wissenschaftliche Veröffentlichungen aus dem Siemens-Konzern*, with contributors from the different Siemens laboratories.

The activities of Hans Riegger and Walter Schottky exemplify Siemens's research agenda in electroacoustics during World War I and its continuity into the interwar period. Schottky worked on amplifying electrical signals, Hans Riegger on their conversion into acoustic signals and vice versa. In the transition from war to peacetime, Riegger transferred practices and instruments from sound transmitters and microphones for underwater acoustics to sound propagation in the atmosphere and the development of loudspeakers and microphones. Riegger's initial research trajectory was quite similar to those of Walter Hahnemann and Heinrich Hecht at Signal Gesellschaft. Like Hahnemann, Riegger had a background in wireless telegraphy and connected oscillation research from electromagnetism with electroacoustics, merging it into a single field of oscillation research (Gerdien 1926, 323). Riegger developed a general theory of electrodynamic sound transmitters, much as Hahnemann and Hecht had. But Siemens was a very different company from Signal Gesellschaft. Whereas Signal Gesellschaft remained a subsidiary of a marine engineering company, after the war, Siemens developed sound recording and large amplification systems that were entirely dislocated from maritime uses. With Riegger's background in underwater sound transducers, he designed a condenser microphone and an electrodynamic loudspeaker that he called the Blatthaller. The Blatthaller had a flat piston membrane, similar to the sound transmitters for underwater telegraphy—but underwater sound transmitters were built to transmit large amounts of sound energy into the water, while the Blatthaller transmitted large amounts of sound into the air.

The development of loudspeakers and microphones at Siemens depended critically on the development of amplifier tubes and amplifiers, which were needed to amplify the weak microphone signals and to drive the loudspeakers. While the sound transmitters for underwater telegraphy operated at a single frequency, supplied by an alternating current from a transformer, the Blatthaller and other loudspeakers needed to reproduce the whole range of audible frequencies in speech and music evenly. The amplifiers not only had to be powerful, but were required to amplify the sound signals in a linear way and with as little distortion as possible. The noise produced by circuits and microphone contacts and the distortion of signals was a major concern; it had already limited the use of microphones and amplifiers during World War I, as discussed in the previous chapter.

In 1912, Siemens and Halske had entered a consortium with AEG and Telefunken to take on the amplifier tube patents of Austrian physicist Robert von Lieben. While Lee De Forest had patented the audion for the amplification of wireless signals, the Lieben patent emphasized the amplification of sound, specifically telephone signals, for long-distance telephony. On the basis of the Lieben patents, Siemens researchers could develop their own amplifier tubes instead of paying license fees to use the American patents. Walter Schottky carried out research on electron currents, which led to the Siemens patent of the screen grid vacuum tube in 1916.[67] His work exemplified the close relationship between the novel theories of relativity and quantum physics and research and development in the electrical industry. Schottky completed his dissertation on energy conservation and dynamics in relativity with Max Planck in Berlin in 1912. He then moved on to experimental investigations of electron currents in high vacuum at Max Wien's institute at the University of Jena. Though prevented by ill health from enrolling in the military, Schottky started working for Siemens after the outbreak of World War I with the aim of supporting the German war effort. Based on his research on electron currents, he set out to study and improve the Lieben tubes. His invention of the screen grid vacuum tube enhanced the performance of Siemens tubes considerably.

At the end of 1917, Schottky started to investigate the electric distortions that limited the amplification of acoustic signals. Through the work of Schottky and other physicists and engineers, the concept of noise, which originated as an exclusively acoustic category describing the quality of a sound, was transformed into something different, describing effects that were inherently electrical. While Barkhausen and others investigated carbon microphones and electric circuits as the source of electric noise, Schottky looked at the electron current of the amplifier tube itself as a source of disturbances that intrinsically limited the amplification of signals. In 1918, he identified the effect that became known as "shot noise" (*Schrotrauschen*) as the lower limit of thermal noise in amplifier tubes (Dörfel and

Hoffmann 2005, 2–5). In his internal report to Siemens in April 1918, Schottky described the irregularities that led to a barely audible noise (Geräusch) (Serchinger 2008, 141). In an article published in 1918 he described the noise produced by this effect as buzzing (Summen) rather than noise (Rauschen) (Schottky 1918, 541). Schottky was describing not the electrothermal distortion itself but the sound it created in the telephone. Listening to the distortion in a telephone or headphone was the only way to physically observe and characterize the distortion. Nonetheless, scientists and engineers quickly took over the notion of electric noise as a way of characterizing the electric effect itself rather than the sound it produced.[68]

In 1918, Siemens hired Erwin Gerlach, who had studied under Otto Lummer and Erich Waetzmann in Breslau, to set up an electroacoustic laboratory at the central laboratory, which was independent of Siemens Research Laboratories and Riegger's work.[69] This created competition among researchers at different laboratories within Siemens and resulted in several very different microphone and loudspeaker designs in the company. Schottky left his position at Siemens in 1919 to start his habilitation (the German postdoctoral degree to qualify for a university teaching position) with Wilhelm Wien at the University of Würzburg. While back at the university, Schottky continued to work for Siemens as a consultant. Schottky and Gerlach set to work on a new loudspeaker design with a light membrane that was not to exhibit any selectivity for certain frequencies or produce any reverberation or upper harmonics. They developed the ribbon loudspeaker, a current-carrying aluminum ribbon, suspended in the gap of a ring-shaped electromagnet, that delivered the desired improvement to the reproduction of speech and music (Schottky 1924; Serchinger 2008, 212–213).

With the ribbon speaker, Gerlach and Schottky introduced the ribbon microphone, which worked according to the same principle. Schottky's contribution was conceptual and theoretical, while Gerlach carried out the actual construction.[70] The envisioned applications for the speaker specifically included sound motion pictures. Work on the loudspeaker itself was more or less finished in 1922, but the amplifier required to drive the speaker needed more work.[71] Schottky and Gerlach presented the speaker at a meeting of the Deutsche Physikalische Gesellschaft in Innsbruck in 1924. The sound quality met expectations, but the ribbons of the speaker proved insufficiently durable; they ripped apart and needed to be replaced frequently. Although the ribbon speaker was abandoned, the ribbon microphone became quite a successful design, produced by Siemens and other companies.

Schottky's work on loudspeakers and microphones led him to an engagement with more foundational problems of acoustics, specifically the relationship of electroacoustics with the acoustics of Helmholtz and Rayleigh. In a 1926 paper in *Zeitschrift für Physik*, he showed how the reciprocity principle of "classical oscillation theory" introduced by Helmholtz in 1859 related to electroacoustics, especially loudspeakers and microphones (Schottky 1926, 689; Trendelenburg 1975, 185). Schottky's final contribution to acoustics was his chapter on electroacoustics in Karl Willy Wagner's volume *Die wissenschaftlichen Grundlagen des Rundfunkempfangs* (The scientific foundations of radio) of 1927.[72]

In 1922, the Siemens physical-chemical laboratory recruited Ferdinand Trendelenburg, who had just finished his dissertation on the thermophone at the Göttingen Institute for Applied Electricity under Max Reich, the successor of Theodor Simon. With Riegger, Trendelenburg worked on the development and analysis of microphones and loudspeakers, on objective tone recording and sound measurement, and on the installation of large amplification systems.[73] Riegger presented his condenser microphone and the Blatthaller at the meeting of the German Physical Society in Innsbruck in September 1924, at the same meeting where Schottky presented the ribbon microphone and loudspeaker. This was Riegger's last public appearance. He was troubled by a severe chronic skin disease, of which he died in March 1926 (Gerdien 1926). The final test for the Blatthaller was the opening of the new premises of the Deutsches Museum in Munich on 7 May 1925. Oskar von Miller, the museum's founder and director, had asked Siemens to set up amplification systems for music and speeches at the sites on Museum Island in the River Isar and on Königsplatz. For Siemens, this was a welcome opportunity not only to test its systems and compare its different microphones and loudspeakers, but also to generate first-rate publicity. With the ribbon microphone by Schottky and Gerlach, the Blatthaller was the core element of the large electroacoustic amplification system that Siemens installed. The ribbon speakers were installed only on the roof of the Propylaea on Königsplatz (figure 4.2), while Blatthaller speakers were used at several other locations as well.[74]

Figure 4.2
Large ribbon speakers on the roof of the Propylaea on the west side of Königsplatz, Munich, at the opening of the Deutsches Museum at its new location on 7 May 1925 (Deutsches Museum, München, Archiv BA-E 0001586).

Following Riegger's death in 1926, Trendelenburg became Siemens's most active acoustician. Trendelenburg authored and edited several textbooks and handbooks (Trendelenburg 1927, 1932, 1934, 1935, 1939). He worked on human speech, speech analysis, and the relationship between objective sound measurement and subjective sound perception (Trendelenburg 1931). In 1929, he completed his habilitation, with a study of the physical properties of heart sounds, at the University of Berlin, where he became professor extraordinarius in 1931 (Trendelenburg 1928).

Siemens continued to develop and test the characteristics of larger speakers with aluminum membranes and an input power of about 800 watts, which could broadcast speech over several kilometers (figure 4.3).[75] After Siemens transferred most of its electroacoustics research to Telefunken in

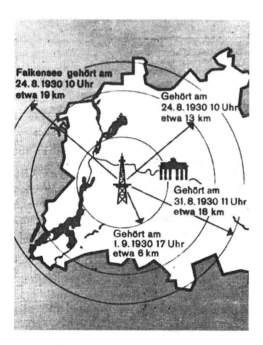

Figure 4.3
Audibility tests with the large Siemens Blatthaller, installed on the Berlin Radio Tower in 1930. The maximum broadcasting distance of the speaker depended crucially on weather conditions. It could be heard up to 19 kilometers away (Trendelenburg 1975, 194).

1931, Trendelenburg did not move to Telefunken, as many of his colleagues did, but worked for a few years in other fields of technical physics. He came back to electroacoustics later in the 1930s, when Siemens recommenced its activities in the field.

Many of the electroacoustics research and development activities at the Siemens laboratories related to the transmission of speech and music through telephone and other circuit networks. Karl Küpfmüller worked with Karl Willy Wagner at the Telegraphen-Versuchsamt (Telegraph Research Office, attached to the Reichspost) from 1919 until 1921, after which he moved on to the Siemens central laboratory. In parallel with Harry Nyquist and Ralph Hartley at Bell Laboratories, Küpfmüller determined the limits of how many telegraph signals, or telephone conversations, could be compressed and transmitted over a telegraph or telephone line while maintaining intelligibility of telegraph signals or telephone speech (Hagemeyer 1979, 167–171). Küpfmüller became professor of electrical engineering at the Technische Hochschule Danzig in 1928 and a professor at the Technische Hochschule Berlin-Charlottenburg in 1935. In 1937, he returned to Siemens to head research and development in electrical communication engineering at the company (Schoen 1994).

While Küpfmüller investigated the transmission of electroacoustic signals on cables, Richard Feldtkeller was working on their distortion-free amplification. Feldtkeller had completed his studies in physics at the University of Halle with a dissertation on parasitic currents in high-frequency amplifiers in 1924, after which he joined Siemens Research Laboratories. Feldtkeller stayed with Siemens until 1937, when he became professor of electrical communication engineering at the Technische Hochschule Stuttgart (Trendelenburg 1975, 165–168; Michel 1994). In 1925, Siemens Research Laboratories hired Ernst Lübcke, who had worked for Atlas Werke between 1920 and 1924. At Siemens, Lübcke continued to work on underwater sound and direction ranging, along with Trendelenburg and Gerdien.[76] Lübcke, Trendelenburg, and Gerdien also took out patents for acoustic echo sounding and direction sensing for aviation.[77] This work on underwater acoustics, echo sounding, and direction sensing for aviation established a continuity with military-related research during World War I, even though its scale was rather small during the Weimar Republic.

Among the various research and development projects in acoustics at Siemens was the Siemens Nernst Bechstein piano, also known as the Neo-Bechstein. In the Neo-Bechstein, the vibrations of the piano strings did not resonate in a soundboard but were picked up by recording coils and electrically amplified. Walther Nernst, then professor of physics at the University of Berlin, developed the Neo-Bechstein in collaboration with Stephan Frankó and his assistants Hans Driescher and Oskar Vierling. The piano manufacturer C. Bechstein built the mechanical parts, while Siemens supplied the electrical parts. As with most electroacoustic music instruments of the interwar period, the Siemens Nernst Bechstein piano did not meet with success, and only a few instruments were built between 1931 and 1933. In September 1932, Siemens transferred its contract with Nernst to Telefunken, as it had done a year earlier with other electroacoustics activities (Barkan 1999, 243–244).[78] But though Siemens divested from amplification systems and electroacoustic musical instruments, it continued to develop electroacoustic measurement equipment. In 1935 Erich Freystedt presented a tone frequency spectrometer developed at the Siemens central laboratory. This was a sound frequency analyzer that visualized the frequencies on the screen of a cathode ray oscilloscope (Freystedt 1935).

4.3.2   Sound Motion Pictures: Acoustics at the AEG Research Institute

The Allgemeine Electricitäts-Gesellschaft (AEG) was Siemens's largest competitor in the German electrical industry. It also produced electroacoustic devices and systems, but electroacoustics was not as central to AEG as it was to Siemens, which had a different company structure. While Siemens and Halske was the main manufacturer and supplier of telegraphy and telephony hardware to the Reichspost, AEG's growth came from high-voltage engineering and large machinery. AEG had aimed to enter the market for wireless telegraphy with the system developed by Adolf Slaby, professor of electrical engineering at the Technische Hochschule Berlin-Charlottenburg, and his assistant Georg Graf von Arco. After the 1903 establishment of Telefunken, which took over wireless from AEG and Siemens, AEG had left research and development in low-voltage engineering largely to Telefunken and Siemens, where Rukop and Schottky researched amplifier tubes during World War I. AEG appears not to have entered the field of underwater acoustics related to submarine warfare either. At the time the Weimar Republic was proclaimed, AEG was conducting little in-house research compared with Siemens.[79] The company wanted that to change. Influenced by the experience with industrial research during World War I, in 1919 Georg Klingenberg took the initiative to establish a scientific research laboratory at AEG.[80] It took until 1928 to set up the Forschungsinstitut der Allgemeinen Electricitäts-Gesellschaft (AEG Research Institute) and start its operations.

The founding of the AEG Research Institute in the Berlin district of Reinickendorf in April 1928 was a major step in strengthening and consolidating AEG's research efforts. AEG appointed Carl Ramsauer, one of Germany's most distinguished experimental physicists, as the institute's director. Ramsauer, formerly a professor at the Technische Hochschule Danzig, was known for his experiments on the scattering of slow electrons by noble gases, which did not follow the laws of classical physics.[81] In the institute's first yearbook, covering 1928–1929, Ramsauer reported that about thirty scientists were working in eight laboratories divided into general physics, general chemistry, acoustics, chemistry of insulating materials, electrical technology, metallurgy, heat technology, and magnetism.[82]

Ramsauer appointed his former student Ernst Brüche to head the physics laboratory, where Ramsauer and Brüche continued investigations into electron scattering and turned their attention to electron optics. Research in electron optics was related to the development of cathode ray tubes, which promised to have a huge potential for television and for electron microscopy. The acoustics research group had a very different focus. It developed a complete system of sound motion picture recording and reproduction for Tobis and Klangfilm GmbH. The research institute workshop even produced the first sound cameras, which were delivered to production companies for film shooting in 1929.[83] To develop a complete system of sound motion pictures was a complex endeavor, as Ramsauer emphasized in 1930.[84] Sound motion technology required research and development not only in acoustics but also in electro-optics, electric amplification, and the chemistry of film photography. The research institute put Hugo Lichte in charge of the acoustics research group. After his deployment at the Torpedoinspektion (Torpedo Inspection Office) in Kiel during World War I, Lichte began working for Signal Gesellschaft in 1919 (Mielert 1985). Following its sale, in 1924 he moved to Mix and Genest AG, an electrical company mainly owned by AEG that produced radio receivers, amplifiers, and loudspeakers. In 1926, Lichte moved on to AEG.

In Germany, attempts to develop a system of sound motion pictures went back to the inventor and film producer Oskar Messter, who presented his Biophon system in 1903. Not until more than twenty years later, in 1927, did the U.S. production company Warner Brothers successfully bring *The Jazz Singer*, the first feature-length sound motion picture, to cinemas.[85] Until *The Jazz Singer* and its follow-up, *The Singing Fool*, the German motion picture and electrical industry was rather reluctant to invest in sound motion pictures, believing that its future was uncertain. The success of Warner Brothers and Western Electric's Vitaphone sound-on-disc system, however, created an urgent need to develop a system that was independent of the American patents. In 1930 Heinrich J. Küchenmeister, one of the entrepreneurs of sound motion pictures in Germany, described the situation:

Inventive genius has worked for decades on the task of making talking pictures; but only the development of technical tools in various fields has brought the problem of sound film to its solution and practical implementation. The rapid developments in the field of radio have advanced the issues of sound film just as much as has the perfecting of electroacoustics in the field of talking machines and gramophone records, and the improvements in film manufacture by the chemical industry. (Küchenmeister 1930, 361)[86]

Although various scientists and entrepreneurs had proposed the basic principles of sound motion pictures independently around the turn of the century, Küchenmeister's words emphasize that no inventor or group of inventors could singlehandedly achieve a breakthrough for the talkies. The success or failure of sound motion pictures cannot be understood in isolation but must be seen in the context of other media technologies, especially radio broadcasting (Jossé 1984, 247–250). As a large technological system, sound motion pictures depended on a wide variety of technological, social, economic, and cultural factors.

Messter's Biophon was a sound-on-disc system. A gramophone coupled to the projector was intended to ensure synchronization between sound and film. Issues of sound quality in recording and reproduction, the lack of adequate sound amplification, and the limited playing times of gramophone records hindered an early breakthrough of sound motion pictures before the war (Jossé 1984, 97–104). The alternative to recording and reproducing sound on phonograph or gramophone discs was to record and reproduce the sound electro-optically on the film itself. The sound-on-film system that Lichte and others eventually developed at AEG resolved the difficulties of synchronization and limited playing time. Though Lichte traced the origins of electro-optical sound recording back to the electric arc designed by William Duddell and Lichte's mentor Hermann Theodor Simon, it was not Duddell nor Simon but Ernst Ruhmer who was credited with having achieved the first electro-optical sound recording and reproduction using an electric arc, in his private Berlin laboratory in 1901.[87]

Ruhmer did not get beyond the experimental stages of electro-optical sound recording before his death in 1913, and the larger problems of sound quality and sound amplification were solved only after World War I. In 1918 the Tri-Ergon group of Vogt, Massolle, and Engl set out to develop a sound motion picture system from scratch, based on optical recording of sound film. But Tri-Ergon did not succeed in transforming sound motion film into a commercially successful system. After Vogt had left his former employer Georg Seibt, Tri-Ergon entered into an agreement with C. Lorenz AG and, in 1925, an agreement to produce sound motion pictures with Universum Film AG (UFA), which had an almost complete monopoly on German film production. UFA's backing was, however, only half hearted. Apart from financial constraints, UFA was not convinced that sound motion pictures would succeed and did not want to commit to the Tri-Ergon system.[88] It was the Warner Brothers movies based on the Vitaphone sound-on-disc system that eventually led to the breakthrough for sound motion pictures in German film theaters.[89] Reacting to strong competition from the Hollywood film studios, which had changed their entire production to talkies, the German film industry founded Tobis—Tonbild-Syndikat AG—in 1928.

UFA, Siemens and Halske, and AEG did not join the Tobis syndicate, having begun to develop their own independent sound-on-film systems. Siemens used a variable area system in which the sound intensity was written in a waveform on the soundtrack, while in AEG's variable density system, the sound intensity was proportional to the transparency of the film. Erwin Gerlach and Fritz Fischer started to work on sound motion pictures at the Siemens central laboratory in 1926.[90] AEG started work on its sound motion picture system in 1927. The company hired Horst Tischner, a former student of Barkhausen's, and Albert Narath, who developed the photochemistry for the AEG system.[91]

For sound recording, the AEG researchers used a condenser microphone and a photoelectric Kerr cell, based on the work of August Karolus, professor of physics at the University of Leipzig, who had worked with Telefunken since 1924. For sound reproduction, the team employed a photoelectric potassium hydride cell. Considerable effort was expended on designing the amplifiers and the optics of the recording and reproduction system in order to keep control of the frequency response, and on the mechanics to secure synchronization. The sound was recorded separately by a special sound camera, then copied together with the images onto a single film for reproduction. The film cameras had to be installed in soundproof boxes.[92]

To consolidate their efforts to develop a competitive sound motion picture system, in 1928 AEG and Siemens founded Klangfilm GmbH, which continued work on the AEG variable density system. In a tense market situation, Tobis and Klangfilm reached an agreement in 1929, with Tobis taking over the production of sound motion pictures and Klangfilm the development and manufacture of film cameras and projectors. Soon thereafter Klangfilm reached a similar agreement with UFA. Tobis and Klangfilm eventually managed to keep the Vitaphone and Western Electric systems out of German cinemas. Other European countries, however, became battlefields for the different systems (figure 4.4). This technological war was settled by the Paris Sound-Film Peace Treaty of 1930, which divided up the international sound film markets between the German and American groups in a cartel-like arrangement.[93]

In 1931, Fischer and Lichte edited a volume in which researchers from AEG and Siemens presented the various aspects of the Klangfilm sound motion picture system. While promoting the sound-on-film system, they

Figure 4.4
"Good sound—good business" and "Klangfilm Europe," two 1930s pamphlets by the German company Klangfilm (Siemens Corporate Archives, file 8353). The international sound motion picture markets became a battlefield for motion picture producers and for the producers of sound equipment, such as Klangfilm, after sound-motion technology experienced its international breakthrough in the late 1920s.

additionally addressed the sound-on-disc recording and reproduction system, which Klangfilm also marketed and installed. As well as explaining the different recording and reproduction apparatus, the book contained a chapter on architectural acoustics, authored by Ferdinand Trendelenburg, and a detailed description of the brand new UFA sound motion picture studios at Neubabelsberg near Berlin.[94]

The activities of the acoustics group at the AEG Research Institute included research on loudspeakers, which had to be developed for the sound amplification systems of cinemas. Heinrich Stenzel worked on the direction characteristics of sound transmitters, the frequency response and acoustic radiation of membranes, and the horn speakers that were eventually installed in sound motion picture theaters.[95] In autumn of 1931, the AEG Research Institute closed down the acoustics laboratory and transferred its sound-motion activities, including a number of the acousticians involved, to the electroacoustics division of Telefunken. Lichte became head of the low-frequency laboratory, where he was in charge of Telefunken's research and development in acoustics, amplifiers, and sound motion pictures, and where he worked with, among others, Erwin Gerlach (formerly of the Siemens central laboratory).[96]

Not all acousticians followed Lichte from AEG to Telefunken. Stenzel took up employment at Atlas Werke in Bremen between 1932 and 1934. From Atlas he went to the German Navy's Torpedoversuchsanstalt (Torpedo Testing Institute), where he headed the division for underwater acoustics.[97] Stenzel was following Lichte's route back to underwater acoustics. This made the dynamics between acoustics research for warfare and new media technologies during peacetime almost reversible. Lichte's work on underwater acoustics with the Torpedo-Versuchs-Kommando during World War I put him on a trajectory to implement sound motion pictures at AEG; in turn, developing loudspeakers for sound film at AEG led Stenzel into underwater acoustics with the Torpedoversuchsanstalt during World War II. Stenzel's move from loudspeakers for sound motion pictures to underwater acoustics for U-boat warfare was by no means coincidental. In the AEG Research Institute's yearbook for 1928–1929, Stenzel emphasized that the directional characteristics of sound transmitters were no less important for underwater sound telegraphy and sound location of submarines and aircraft than for radio and sound motion pictures. In fact, Stenzel's involvement in underwater acoustics itself had a prehistory. Before becoming part of the AEG Research Institute's acoustics group, Stenzel had published on echo sounding. In January 1928, with Heinrich Hecht, he filed a patent for an "arrangement for directional transmission and reception of sound" for Electroacoustic GmbH.[98]

Like Siemens, AEG did not turn all its electroacoustics activities over to Telefunken. It began to develop magnetic tape recording in 1932, after reaching an agreement with Fritz Pfleumer, the inventor of the principle. Pfleumer embedded steel powder in paper or synthetic tape that he then magnetized using a recording head. Magnetic tape recording combined the advantages of optical sound-on-film with magnetic recording on steel tape or wire. Both optical sound-on-film and magnetic recording were used in radio broadcast studios in the early 1930s. Sound-on-film was lightweight and made it possible to cut and edit sound in the same manner as motion pictures. Recordings on steel tape or wire could be erased and the medium reused, but steel tape or wire was heavy and could not be cut and spliced. In contrast to sound motion pictures earlier on, activities in magnetic tape recording did not take hold in the AEG Research Institute. AEG hired the electrical engineer Eduard Schüller, who had written his diploma thesis on magnetic recording on wire at the Heinrich Hertz Institute in 1931. The main difficulty in making magnetic tape recording a reliable device was the mechanical stability of the tapes, which were made by BASF. Most issues were overcome when BASF and AEG began to use magnetite powder in plastic tapes instead of the coarse carbonyl iron they had previously employed in paper tapes. AEG introduced the first magnetic tape recorder at the Berlin Radio Show in 1935 (F. K. Engel 1999; Schüller 1965).

With the transfer of most electroacoustics activities from Siemens and AEG to Telefunken in 1931, Telefunken set up its own department of electroacoustics, the Elektroakustische Abteilung, generally referred to as Ela. Telefunken developed and built microphones, loudspeakers, and large electroacoustic amplification systems for open spaces. These amplification systems soon found a huge market with the takeover of the German state by Adolf Hitler and the Nazi Party (NSDAP) in 1933, which had a large appetite for mass public orations and spectacle, a topic addressed in chapter 5.

4.4 Public Research Laboratories

So far we have looked at the Technische Hochschulen, the universities, and the electrical industry as settings for acoustics research and development in the interwar period. But starting from the end of the nineteenth century, a whole range of new institutions appeared in the German research landscape, which were positioned between academia, the state, and industry. The prototypes of these public research institutions were the Physikalisch-Technische Reichsanstalt and the Kaiser Wilhelm institutes, introduced in previous chapters.[99] The Physikalisch-Technische Reichsanstalt, founded as a national research and testing laboratory for German science and industry in 1887, held a special position among public research institutions in Germany and served as a model for similar institutions, such as the National Bureau of Standards in the United States and the National Physical Laboratory in the United Kingdom. While the National Physical Laboratory quickly became an important institution for acoustics research in Britain, the Physikalisch-Technische Reichsanstalt did not establish its acoustics laboratory until 1934. Neither did the Kaiser Wilhelm institutes carry out significant acoustics research during the Weimar Republic.

In the 1920s voices in the physics community were already calling for the Reichsanstalt to become involved in the new acoustics research. Max Wien, one of the Reichsanstalt trustees, had argued repeatedly for an independent department of acoustics. Wien envisioned that the department would be active in fields like architectural acoustics, the testing of loudspeakers, and medical acoustics. Friedrich Paschen, president of the Reichsanstalt, countered that such a department was not necessary, since the establishment of a new major research institution was on its way. The Heinrich Hertz Institute for Oscillation Research would carry out exactly the kind of research and testing activities that Max Wien contemplated, and the German Reich could not allow itself to fund two institutions that would duplicate each other's work.[100]

The Heinrich Hertz Institute opened its doors in 1930. Its first director, Karl Willy Wagner, had been carefully orchestrating the foundation of the new institute since 1926. Like Barkhausen and Lichte, Wagner had earned his doctorate with Theodor Simon at the Institute for Applied Electricity of the University of Göttingen before the war, and he adopted Barkhausen's program of Schwingungsforschung, bringing together electrical, acoustical, and mechanical oscillations, as the organizational blueprint and research program for the Heinrich Hertz Institute. This program, along with close relationships with the electrical industry, placed the institute in a unique position to become a major player in the new world of electroacoustics research.

While the Heinrich Hertz Institute was unquestionably the most important public research institution for acoustics and electroacoustics in interwar Germany, it was not the only one. The Rundfunkversuchsstelle (Radio Experimental Office), which opened at Berlin's School of Music in 1928, was another, although much smaller, facility. The Rundfunkversuchsstelle was established to experiment with the new sounds and to create electroacoustic musical instruments and artistic work related to electroacoustic media technologies, especially radio broadcasting and sound motion pictures. At the Rundfunkversuchsstelle, Friedrich Trautwein and Oskar Sala worked on the trautonium, one of the most-discussed electroacoustic musical instruments of the interwar period.[101] The Institut für Schall- und Wärmeforschung (Institute for Sound and Heat Research) of the Technische Hochschule Stuttgart was a public research institute active in acoustics research similar in structure to the Heinrich Hertz Institute. The following section takes a closer look at acoustics research, both at the Heinrich Hertz Institute and at the Institut für Schall- und Wärmeforschung.

4.4.1   The Heinrich-Hertz-Institut für Schwingungsforschung

The early history of the Heinrich Hertz Institute for Oscillation Research is inextricable from its founder Karl Willy Wagner. Wagner graduated as an electrical engineer from a small technical college and worked at the heavy-current laboratory of Siemens Schuckertwerke. In 1908, his interest in the scientific aspects of electrical engineering led him to Theodor Simon at the Institute for Applied Electricity in Göttingen, where he obtained his doctorate with a dissertation on the electric arc as a generator for alternating currents (Wagner 1910). Wagner then investigated oscillations and the propagation of waves in electrical systems at the Telegraphen-Versuchsamt, and he completed his habilitation at the Technische Hochschule Berlin-Charlottenburg in 1912. He joined the Physikalisch-Technische Reichsanstalt in 1913. Wagner worked on several military projects during World War I, among them a method for secret telephony and a project to guide submarines through minefields, as well as supervising the development of radio transmitters for aircraft at Telefunken.[102] Finally, Wagner became president of the Telegraphentechnisches Reichsamt in Berlin in 1923.

The Telegraphentechnisches Reichsamt had been established in 1920 to coordinate the Reichspost's various test laboratories. Since its investigations were always directly related to technical tasks of the Reichspost, it never took on the character of a scientific research institution (Hagemeyer 1979, 465). Wagner, however, was determined to establish a solid research infrastructure for state-run media technologies. In 1924, he founded a new journal, *Elektrische Nachrichten-Technik* (Electrical communications technology), which became a central organ in its field.[103] In the same year, Wagner initiated the Heinrich Hertz Gesellschaft zur Förderung des Funkwesens (Heinrich Hertz Society for the Advancement of Radio) to support radio research. The society organized a series of lectures in 1925 and 1926, where expert scientists from universities, Technische Hochschulen, corporate laboratories, and the Reichspost presented the different scientific aspects of radio.

In 1927 Wagner published *Die wissenschaftlichen Grundlagen des Rundfunkempfangs*, an edited book that contained some of the presentations of the lecture series. In the introduction, he stressed the rapidly growing cultural and economic importance of radio in Germany and in an international context. The various chapters revealed the inherently electroacoustic character of radio, presenting the acoustics of radio as inseparable from its electrical operation on all levels. The list of contributors to *Die wissenschaftlichen Grundlagen des Rundfunkempfangs* included Franz Aigner, Walter Hahnemann and Heinrich Hecht, Walter Schottky, Hans Rukop, and Heinrich Barkhausen. Aigner discussed the vibrations of speech and music and their reproduction and distortion in radio. Hahnemann (now at C. Lorenz AG) and Hecht presented their electroanalog conception of the sound field and electroacoustic transmitters, which derived from their work on underwater sound telegraphy but was equally applicable to radio. In the chapter entitled "Elektroakustik," Schottky discussed the theory and the characteristics of loudspeakers and microphones, and presented his work on reciprocity in electroacoustic transducers. Rukop and Barkhausen contributed the chapters on amplifier tubes and amplifier theory (Wagner 1927).

On Wagner's initiative, in 1926 the German Reichspost, the Prussian Ministry of Science and Education, and the Reichs-Rundfunk-Gesellschaft set up a plan to establish a scientific research institute for electric and acoustic vibrations and oscillations. Wagner then took charge of founding the Studiengemeinschaft für Schwingungsforschung (Study Group for Oscillation Research), which became the patron of the Heinrich Hertz Institute. It had representatives from the Reichspost, the electrical industry, and the Technische Hochschule Berlin-Charlottenburg. The electrical industry contributed 450,000 reichsmarks to the institute; 200,000 reichsmarks came from the Reichsbahn (the state-owned railway) and related industry; while the Reichspost, the Ministry of Culture, and the National Broadcasting Corporation covered the annual running costs of 185,000 reichsmarks. The institute was affiliated with the Technische Hochschule Berlin-Charlottenburg, which appointed Wagner as professor of *Schwingungslehre* (oscillation studies) in 1927.

The Telegraphentechnisches Reichsamt, renamed the Reichspostzentralamt (Central Office of the Reichspost) in 1928, organized three- to four-day seminars each year between 1926 and 1929, bringing together professors from the Technische Hochschulen and universities with representatives of the Reichsamt. The presentations included topics from acoustics. A small number of delegates from the Reichswaffenamt or the Heereswaffenamt (Army Weapons Office) attended the meetings, while the electrical industry seems to have been excluded.[104] The Heinrich Hertz Institute for Oscillation Research celebrated its opening in March 1930 with five departments: heavy-current technology, telegraphy and telephony, high-frequency technology, acoustics, and mechanics. The heads of the five departments were simultaneously adjunct lecturers or professors (nichtbeamtete außerordentliche Professoren) at the Technische Hochschule Berlin-Charlottenburg. A significant number of the close to fifty temporary scientific staff were advanced students and doctoral candidates.[105]

The establishment of the Heinrich Hertz Institute was explicitly framed as an attempt to catch up and compete with the U.S. research community and, consequently, the U.S. electrical industry. Yet, the institute was not a copy of the structures and practices of American research facilities; in fact, Wagner described it as a unique institution.[106] Its task was to bring together the different fields of oscillation research, such as mechanical vibrations, electromagnetic oscillations, and acoustics. At the same time, the institute was to bridge the gap between the electrical industry, public telecommunications services, and academic research. Even though many of the corporate laboratories already had close ties with the University of Berlin and the Technische Hochschule Berlin-Charlottenburg, the Heinrich Hertz Institute could establish a different research agenda that went beyond the interests of a single corporate laboratory or academic department.

Wagner appointed Erwin Meyer to head the institute's Department of Acoustics in April 1929. Meyer's biography shared many aspects with that of Ferdinand Trendelenburg.[107] After participating in World War I as a soldier, Meyer studied physics at the University of Breslau and completed his dissertation under Waetzmann in 1922 with a study of the forces exerted on membranes by sound waves.[108] Wagner recruited Meyer to the Telegraphentechnisches Reichsamt in 1924 as an acoustics researcher. At the Reichsamt, Meyer investigated the acoustic features of telephony and radio broadcasting, especially the authentic reproduction of music. He worked on various aspects of loudspeakers, microphones, human hearing, and architectural acoustics. Many of his publications were on electroacoustic sound measurement. Meyer proposed a method of analyzing compound tones using a carbon microphone, a Wheatstone bridge circuit, and a galvanometer.[109] This became the topic of his habilitation thesis at the Technische Hochschule Berlin-Charlottenburg in 1928. He was appointed professor extraordinarius in acoustics in 1934, and full professor in 1938. Like Wagner and the other heads of departments at the Heinrich Hertz Institute, Meyer was appointed not in technical physics but at the faculty of machine engineering.

The research of the Department of Acoustics at the Heinrich Hertz Institute covered a broad range of subjects, including sound measurement, architectural acoustics, electroacoustic devices, sound insulation and noise abatement, and musical acoustics. Meyer soon became one of Germany's leading acousticians, with an international reputation—he gave invited lectures in the Soviet Union in 1931, the United States in 1936, and Britain in 1937.[110] The Department of Acoustics was not the only actor in electroacoustics-related research at the institute. This was part of Wagner's agenda of using overarching oscillation research to unite the field of acoustics with high frequency, radio, and telephony and to break down the boundaries between them. Wagner himself engaged in several acoustic research projects in the late 1920s and early 1930s, especially related to noise abatement and the electrical analysis and synthesis of human speech (see figure 4.5). He carried out this work in his own laboratory at the institute, the "director's laboratory."[111]

Even though music was clearly not the main reference for acoustics research at the Heinrich Hertz Institute, the institute was a main player in the field of electric music and the development of new electroacoustic music instruments, for which Berlin had become a center (Donhauser 2007, 67–126). Erwin Meyer investigated the sound spectra of musical instruments (Meyer 1931). It was, however, not Meyer's Department of Acoustics but Gustav Leithäuser, head of the Department of High-Frequency Technology, and Wagner's assistant Oskar Vierling, who spearheaded the development of new electroacoustic musical instruments at the institute. Leithäuser was a board member and honorary president of the Gesellschaft für elektrische Musik (Society for Electric Music), which was founded in Berlin in 1932 and had close ties with the

Figure 4.5
Karl Willy Wagner and his laboratory setup for the electroacoustic vocal synthesizer at the Heinrich Hertz Institute, Berlin. The photo is not dated but was probably taken in the mid-1930s (Archiv der Berlin-Brandenburgischen Akademie der Wissenschaften: NL K.-W. Wagner).

Rundfunkversuchsstelle.[112] Leithäuser was also the spokesman of the institute's Electronic Orchestra, which performed to great acclaim at the Berlin Radio Show in 1932 and 1933.[113]

Oskar Vierling was the most active in the development of new electroacoustic musical instruments at the institute. He had begun his career as an engineer at the Telegraphentechnisches Reichsamt in 1925, where he had collaborated with Jörg Mager, a pioneer in electroacoustic music (Donhauser 2007, 23–42). Vierling participated in the development of the Neo-Bechstein piano as Nernst's assistant, then became Wagner's assistant at the Heinrich Hertz Institute. There he developed an electroacoustic piano, which he named Elektrochord, and a number of electrified string instruments, such as the electric violin and the electric cello, where the soundboard was replaced with electrostatic pickups (102–111). Vierling's instruments became part of the Electronic Orchestra, together with the trautonium and the Neo-Bechstein piano. Vierling obtained his doctorate in 1935 with a thesis on the electroacoustic piano, which was produced by the Förster Piano Factory. Like most of the electroacoustic musical instruments of the interwar period, the Elektrochord was not an economic success, and few instruments were produced.[114]

By 1938, the Department of Acoustics employed nine scientific staff besides its head Erwin Meyer, four doctoral students, and three technicians.[115] Much had changed at the institute by then, after Heinrich Fassbender took on its leadership in 1936. Despite its worldwide recognition, Heinrich Hertz's name was dropped for political and ideological reasons—Hertz had been classified as "non-Aryan." Karl Willy Wagner, Gustav Leithäuser, and others were removed for personal, political, and ideological reasons. Military relevance slowly but steadily entered the research agenda of the Department of Acoustics and the entire institute (see chapter 5).

4.4.2   The Institut für Schall- und Wärmeforschung

The Institut für Schall- und Wärmeforschung (Institute for Sound and Heat Research) at the Technische Hochschule Stuttgart was founded in 1929. The idea of the institute went back to Oscar Knoblauch's Laboratory for Technical Physics at the Technische Hochschule Munich. The building trade was in transformation in the first decades of the twentieth century, with craft-based construction being replaced by industrial practices and laboratory-based engineering. Construction materials were increasingly standardized, tested, and certified by materials-testing laboratories attached to the Technische Hochschulen. Changing building practices and the acoustic adaptation of motion picture theaters and other premises for sound amplification created a market for consulting, testing, and certifying novel construction materials and methods in terms of their sound-absorbent properties (E. Thompson 2002, 190–207; Wittje 2003, 118–120). Knoblauch had initiated the Forschungsheim für Wärmeschutz (Research Facility for Heat Insulation) in Munich in 1918, and this served as a blue-print for what Knoblauch had in mind for sound and heat research in Stuttgart.

The plan to found an institute for sound and heat research was not envi-sioned for the national level but was a regional initiative for the state of Württemberg, where Stuttgart and its Technische Hochschule were located. Initial discussions between the Württemberg Ministry of Culture and repre-sentatives of the Technische Hochschule Stuttgart about establishing an institute for sound and heat research to serve the building industry took place in 1927. Julius von Jehle, president of the state of Württemberg's Landesgewerbeamt (Trade Office), entered the debate in 1928. Knoblauch wanted the Stuttgart institute to be independent of the Technische Hoch-schule, but his standpoint did not prevail. The Ministry of Culture and representatives of the Technische Hochschule Stuttgart insisted on an affili-ation with the Technische Hochschule. In the end, the Institute for Sound and Heat Research was founded as an independent division of the Materials Testing Institute at the Technische Hochschule. Knoblauch passed the reins to the first director of the new institute, his former assistant Hermann Reiher, from whom he seems to have become somewhat estranged in the aftermath.[116]

Knoblauch had appointed Reiher as his assistant after Reiher's gradua-tion with a diploma in electrical engineering from the Technische Hoch-schule Munich in 1922. Reiher completed his doctoral dissertation in 1924 and his habilitation in 1926, after which he became a Privatdozent (adjunct professor) of technical physics at Munich. Reiher's doctoral research was on heat transfer, his habilitation on sound insulation in buildings (Knob-lauch 1941–1942, 34–35; Reiher 1932c)—a combination that made him the ideal director for the Institute for Sound and Heat Research. Richard Heilner of Deutsche Linoleum Werke in Bietigheim, close to Stuttgart, became director of the Verein zur Förderung der Anstalt für Schall- und Wärmetechnik (Association for the Advancement of the Institute for Sound and Heat Technology), which sponsored the institute and secured support from the regional construction industry.[117] Reiher and Heilner argued that the institute should address both physical theory and civil engineering practice. They estimated the initial cost for furniture and equipment at 65,000 reichsmarks and annual running costs at 29,000 reichsmarks, of which 20,000 reichsmarks would be spent on salaries for the scientific and technical staff, including the director, two assistants, and one to two technicians. Consultancy and testing were to secure the institute's funding.

When the Great Depression hit, the German construction industry was unable to keep up its financial commitment to the Institute for Sound and Heat Research. Financial support now came from several foundations, among them the Helmholtz Foundation and the Notgemeinschaft der Deutschen Wissenschaft. In the beginning, the institute occupied two rooms of the Stuttgart Materials Testing Institute and barracks belonging to the city of Stuttgart. The institute's own laboratory building was constructed

Figure 4.6
The north side of the Institute for Sound and Heat Research laboratory building under construction, c. 1930. Most of the construction materials were donated or part of ongoing test series on sound and heat insulation. The building was a patchwork of sixty different outer wall segments, twenty-five different ceilings, forty-five different inner walls, six different roofings, and twenty-four different windows (Reiher 1932a). It was torn down in 1937 to make way for the Reich Horticultural Exhibition of 1939 (Photo: Fraunhofer IBP).

in 1931—mainly with test samples and donations of construction materials from local industry (figure 4.6), which kept the construction budget down to just 25,000 reichsmarks. It included a large reverberation room and a smaller, highly absorbent room for sound measurements, similar to rooms in the Heinrich Hertz Institute in Berlin (Reiher 1932a, 1932b). The Institute for Sound and Heat Research combined applied research in acoustics and thermodynamics, contract consultancy, and the testing of construction materials and methods for their sound- and heat-insulating properties. Research and consultancy included sound measurement and sound abatement in water tubes and ventilation systems. The institute also studied mechanical vibrations that were not necessarily within the audible range but that could damage the structure of buildings.

Reiher and his assistants measured reverberation time in sound motion picture theaters, concert houses, and lecture halls, and made recommendations on how the acoustics could be improved depending on the use of the premises (Stumpp 1936). Special sound-absorbing materials and construction designs were tested in the laboratory and in field experiments in experimental housing and hospitals. The Stuttgart acousticians investigated how airborne sound traveled through walls, and measured impact sound originating from the floor above (Reiher 1932c; Bausch 1939). Despite new materials and scientific testing, the sound insulation of new constructions in the interwar years was not necessarily superior to prewar buildings. On the contrary, the economic crisis of the 1920s and 1930s, economizing measures in the building industry, and the abandonment of craft culture in favor of industrialized production and construction had in many cases led to poorer quality buildings. New, lighter materials and construction methods tested for heat insulation and structural stability did not necessarily have desirable acoustic properties. Reinforced concrete, which was increasingly used in construction, turned out to be a very good conductor of sound vibrations (Wittje 2003, 120; E. Thompson 2002, 209–210; Bausch 1939, 3).

The measurement instruments that Reiher and his assistants used were purely electrical, including microphones, valve amplifiers, tone generators, filters, and decibel meters. All measurements were automated: a standardized hammer created impact sounds; an electric wail tone generator created airborne sounds. Interestingly, Reiher often referred to the work of scientists during World War I, for example, specifically crediting William Tucker's work when he employed hot-wire microphones (Reiher 1932c,

15). Reiher had graduated in electrical engineering, and he exhibited a predominantly electroacoustic understanding of sound in civil engineering. He used electroacoustic analogies and adopted the notion of the sound field with explicit reference to Hahnemann and Hecht's article of 1916 (Reiher 1932c, 7–8; see also Bausch 1939, 4). When explaining the transmission of sound through a wall in a building, Reiher characterized the wall as a membrane, in analogy with the piston membrane of Hahnemann and Hecht's sound transmitters for underwater telegraphy (Reiher 1932c, 10).

The discussions and references cited in Reiher's and his assistants' publications include both German and American literature, demonstrating the international character of acoustics at the time. Yet, Reiher and his assistants did not publish in journals that were likely to be read by other acousticians. Most of their publications appeared in *Gesundheitsingenieur* (Health engineer), an engineering journal that focused on building technology, indoor climate, and health quality. While the Stuttgart acousticians placed their work within international standards of acoustic knowledge and practices, their own testing and consulting had a local and regional focus, and the audience for their publications consisted mainly of civil engineers and architects. The Institute for Sound and Heat Research did not attempt to be a national institution like the Heinrich Hertz Institute.[118]

The leadership of the Technische Hochschule Stuttgart and the Ministry of Culture had insisted that the Institute for Sound and Heat Research would be affiliated with the Technische Hochschule. Despite these efforts, the institute was never integrated into the Hochschule. This had severe consequences for Reiher, who was not allowed to teach in Stuttgart and had to teach in Munich in order to keep his status as Privatdozent. Architectural acoustics for architects, a subject that was clearly Reiher's, was taught by the Stuttgart Privatdozent and theoretical physicist Werner Braunbek in the academic year of 1929–30.[119] Students from the Technische Hochschule Stuttgart could not work at Reiher's institute. In view of these problems, in 1932 the Association for the Advancement of the Institute for Sound and Heat Technology attempted to have the institute integrated into the Technische Hochschule Stuttgart as a separate department.

Reiher, who had taken his habilitation in technical physics in Munich, wanted the Institute for Sound and Heat Research to be a department of technical physics at the Technische Hochschule. Erich Regener, professor of physics at the Technische Hochschule Stuttgart, disagreed. Stuttgart already had its own curriculum in technical physics, in which students took engineering classes in addition to their physics classes, but Regener did not support a separate professorship in technical physics. Neither did he see Reiher as qualified. Regener considered Reiher's research on sound and heat insulation too narrow and his investigations merely practical, believing that Reiher lacked interest in studying the physical foundations of the phenomena.[120] In the academic year 1933–34, Reiher was finally able to teach "Technisch-physikalische Probleme im Bauwesen" (Technical and physical problems in civil engineering) in Stuttgart, including sound insulation and architectural acoustics, but at the faculty of architecture, not that of science.[121] Reiher was appointed professor of technical physics in Stuttgart in 1935, when Regener was forced out in the wake of the Nazi takeover. Reiher could now make the Institute for Sound and Heat Research into the Technische Hochschule's Department of Technical Physics. He lost his professorship again with the denazification process of 1945–1947 and Regener's return (see chapter 5).

In many ways, the Institute for Sound and Heat Research was similar to the Heinrich Hertz Institute. Both were founded with government sponsorship, located somewhere between a Technische Hochschule and industry, and funded by an association that included industry representatives. Both were established in the late 1920s, before the Great Depression. Both combined technical physics, engineering, and testing and consultancy for industry. For both institutes, acoustic research, measurement, and consulting were part of a broader agenda. And both experienced friction in their relationships with the Technische Hochschule to which they each were affiliated. But there were also important differences. Karl Willy Wagner, founder and director of the Heinrich Hertz Institute, had a powerful position within the German scientific and engineering community. His institute was located in Berlin, at the center of German political and industrial power. It was explicitly intended to be an institute of national importance that competed on the international stage. Its program was related to electrical engineering, which in turn was closely affiliated with physics. It cooperated with the electrical and radio industry and the radio broadcasting establishment. In contrast, Hermann Reiher of the Institute for Sound and Heat Research was a little-known Privatdozent from Munich. The Institute for Sound and Heat Research was also much smaller than the Heinrich Hertz Institute. While Stuttgart and its surroundings had significant industries, it was located on the periphery of the German Reich. The institute's agenda was related to civil engineering, which was generally further away from physics than was electrical engineering.[122] It cooperated mainly with the local construction and construction materials industries. And though the research and testing carried out by the Institute for Sound and Heat Research was very useful to the building industry and to the development of sound abatement and sound insulation in civil engineering, its affiliation with civil engineering was a clear threat to its standing within the physics community.

4.5   Institutions of Technische Akustik in the Weimar Republic

The geographical distribution of institutions engaged in acoustics research in Weimar Germany shows an important feature: whereas universities and Technische Hochschulen were rather homogeneously spread over Germany, corporate research laboratories and public research institutes were concentrated in Berlin. The reorganization of the soundscape in the interwar period was predominantly an urban phenomenon. Urban housing construction and city planning were changing, and noise abatement became an issue in large cities. New construction materials, standards, and methods had to be developed and tested. Acoustics research was carried out most intensively in the research laboratories of the electrical industry, the federal radio broadcasting agencies, and the German film industry, all based in Berlin. This high concentration of research institutions made Berlin the undisputed center for acoustics research in Germany.

In fact, the lines between the different kinds of institutions were not drawn sharply. The leaders of public research institutes and corporate laboratories were honorary or adjunct professors at either a university or a Technische Hochschule; they supervised students and published scientific articles and monographs.[123] Researchers at earlier stages in their careers defended their doctoral theses or habilitationen at a Technische Hochschule or university while working at a corporate laboratory or public research institute. Nevertheless, the relationship between regular professors at universities and researchers at other institutions was often tense. Scientists at corporate or public laboratories were frequently seen as too narrow in their research focus and too inexperienced as teachers to be appointed to professorships, especially in physics, as we have seen in the cases of Hermann Reiher, Ferdinand Trendelenburg, and Erwin Meyer.

One of the challenges for technical physics as a community was its fragmentation into a number of subdisciplines, such as technical mechanics, technical optics, technical thermodynamics, and technical acoustics. Acousticians faced three alternatives: identifying with technical physics, identifying with the equally fast-growing community of electrical communication engineering, or following the example of the Acoustical Society of America to create their own representation. All three strategies were followed in interwar Germany. A survey of the first issues of *Zeitschrift für technische Physik* shows many articles on acoustics research related to warfare, which the authors were unable to publish during the war. By the second half of the 1920s, a rapidly increasing number of articles related to electroacoustic devices and media technologies had taken over the agenda of acoustics. Articles on technical acoustics and electroacoustics were also published in general physics journals—including *Annalen der Physik*, *Zeitschrift für Physik*, and *Physikalische Zeitschrift* —and in the general science journal *Naturwissenschaften*. Electrical engineering journals, specializing in radio, electrical communication, and low-voltage engineering, published articles on electroacoustics. Prominent among these was *Elektrische Nachrichten-Technik*.

The growing importance of technical acoustics and electroacoustics started to affect the market for scientific and engineering journals. In 1931, the editors of *Zeitschrift für Hochfrequenztechnik*, Jonathan Zenneck and Eberhard Mauz, changed its name to *Zeitschrift für Hochfrequenztechnik und Elektroakustik*, reflecting the greater role of electroacoustics within physics and electrical engineering. In 1928, Richard Berger founded *Die Schalltechnik*, the first journal entirely dedicated to acoustics. *Die Schalltechnik* was more a technical than a scientific journal, and scientists continued to publish their acoustics research elsewhere. The first academic journal devoted to acoustics alone was *Akustische Zeitschrift*, founded in 1936.

By then, Germany was three years into its new regime, after the takeover of Adolf Hitler and the NSDAP in 1933. In these three years, the new regime had a profound effect on the practice and organization of science in Germany. That effect was especially strong for acoustics research, which was connected to the NSDAP's interests and ideology at several levels. Unsurprisingly, Johannes Stark, the most eminent and outspoken representative of the anti-Semitic movement known as Deutsche Physik, initiated the new *Akustische Zeitschrift*.[124] Stark had never worked in the field of acoustics himself, but in 1934 he founded a new acoustic laboratory at the Physika- lisch-Technische Reichsanstalt, after becoming the Reichsanstalt president in 1933. The field of technical acoustics corresponded to Stark's image of an Aryan physics, in line with Deutsche Physik. Acoustics was applied and industry oriented; its theory was entirely based on classical mechanics; and above all, it was highly relevant for Germany's rapid remilitarization and war preparations (D. Hoffmann 1993). Stark appointed Martin Grütz- macher as the head of the Reichsanstalt's acoustical laboratory and as one of the *Akustische Zeitschrift*'s editors, with Erwin Meyer of the Institute for Oscillation Research—now stripped of its affix "Heinrich Hertz." These changes at the Physikalisch-Technische Reichsanstalt and at the Heinrich Hertz Institute were symptomatic of the impact of the Nazi takeover on the German acoustics community.

5   Acoustics Goes Back to War: Mass Mobilization and Remilitarization of Acoustics Research

5.1   A New Sound for a New Time? Acoustics and Nazi Germany

On 30 January 1933, Adolf Hitler was sworn in as chancellor of Germany by President Paul von Hindenburg. This date marks the National Socialist *Machtergreifung* (seizure of power) and the end of the politically and economically unstable Weimar Republic. In this chapter, I explore the connections between acoustics, the attempt of the NSDAP (National Socialist German Workers Party) to control various soundscapes, and its ideology and practices. What did the seizure of power by National Socialists (NS) mean for acoustics as a scientific and technological field?

The NS seizure of power had a profound effect on both the acoustics research community and the kind of acoustics research carried out. Jewish and politically unwanted scientists were removed while others aligned themselves with the new power structures and advanced their careers. There was an intersection of acoustics research and development with parts of Nazi ideology. Acoustics in the 1930s was largely applied science and not associated with the abstract theories of quantum mechanics and general relativity, which were attacked by representatives of Deutsche Physik (Aryan physics). But for the NS, more important than racist ideology was the fact that acoustic knowledge and practices were useful. Within NS ideology, science had no value as a knowledge-seeking activity in its own right. Like other cultural activities, it had to be made subject and useful to the larger cause of National Socialism.

Acoustics was an important component of motion picture and radio broadcasting, which was among the first institutions the NS brought under its control after it seized power. From the beginning of the political movement, the NSDAP attempted to establish an acoustic presence in urban spaces, a presence it understood as an affirmative resource for social imagination (Birdsall 2012). The takeover and *Gleichschaltung* (political alignment) of mass media by the Ministry of Propaganda under Joseph Goebbels was central to the NS redefinition and reorientation of the German nation. Charlie Chaplin, probably the greatest contemporary satirist of Adolf Hitler, did not miss the importance of the microphone and the soundscape to the Nazi propaganda apparatus in his parody *The Great Dictator* of 1940 (figure 5.1). We should not be misled, however, to overstereotype or mystify the NS soundscape. The NSDAP adapted already existing sound and media technologies to orchestrate its acoustic presence. But other political actors did so as well. The NS use of electroacoustic technologies has to be seen in the context of the use of these technologies by other groups and in other countries (Wijfjes 2014).

The use of electroacoustic technologies during the Third Reich had specific traits. The best known among these is the design and dissemination of the *Volksempfänger*, a cheap mass-produced radio receiver. Moving beyond private homes, the NS developed and installed *Gemeinschaftsempfang* (communal radio broadcasting) systems in schools, factories, offices, and public spaces. Large amplification systems were developed and used to orchestrate

Figure 5.1
Charlie Chaplin as Adenoid Hynkel in *The Great Dictator*, 1940 (Photo: *The Great Dictator*, © Roy Export S.A.S. Scan Courtesy Cineteca di Bologna).

political rallies for the masses and propaganda shows like the 1936 Olympic Games in Berlin. These large amplification systems and the need to control amplified speech and music in vast open spaces put specific demands on the acoustic design of the system and its components (Donhauser 2007, 160–168).

With the remilitarization of Germany in the 1930s, equipping public spaces with sound amplification systems had to compete with the emerging war economy and its constraints on the production capacities of the electrical industry. Acousticians and other actors within the NS movement revisited the experiences of World War I and directed acoustics research again toward aircraft location and U-boat warfare. While acoustic aircraft location was largely replaced by radar during the 1930s, acoustics remained highly relevant for submarine warfare during World War II when it came to U-boat detection as well as acoustic guidance and fuses for torpedoes.

5.2 "Technik ist Dienst am Volke"—Acoustics as Ideology

"Technik ist Dienst am Volke!" (Technology is service to the [German] people!) was the title of Walter Parey's front-page article of 6 July 1935, when he reported on the annual meeting of the Verein Deutscher Ingenieure (VDI; Association of German Engineers).[1] Parey's title can be taken as an appropriate description of the NS approach to science and technology. "Volk" in Nazi jargon always meant the Volksgemeinschaft, an imagined community defined by explicitly racial criteria, and never individual people. The individual, like science and technology, had to be subordinated to the interests and objectives of the Volksgemeinschaft, which was expressed by its leaders. As NS educationist and rector of Frankfurt University Ernst Krieck expressed, "[T]he age of 'pure reason', of a science 'free of prerequisites' and 'free of values' is over."[2] The scientific enterprise had no merit in itself but was expected to serve National Socialism.

Acoustics was a scientific field that had much to offer the NS in creating the acoustic presence of the NS state. Large amplification systems, for which Telefunken was the main producer, were designed and used to lead the masses. In the brochure *Großübertragungsanlagen im Dienste der Volksführung* (Large amplification systems in the service of the people's leadership) of 1938, Telefunken reprinted a passage from the NS radio journal *NS-Funk*:

In the New Germany, radio technology is not a science that gives material to a few insiders to broaden their knowledge and to play around with it by means of technology. To the contrary: Since the takeover of power by the national socialist government, radio broadcasting has been carried into the nation. It became the mediator between state leadership and the population. It became the direct *Erlebnisgestalter* [designer of the experience] of the grand celebrations of the German people.[3]

NS radio and large amplification systems were not designed simply to convey a message; they were meant to create emotional experiences, the soundtrack of the age of National Socialism. NS propaganda could also mobilize acoustics in less presumptuous ways. While large amplification systems broadcast with hitherto unknown sound intensities, the NS Reichsgemeinschaft der technisch-wissenschaftlichen Arbeit (RTA; Society for Technical and Scientific Labour) and the Deutsche Arbeitsfront (DA; German Labour Front) adopted the mission of noise abatement and organized the Reichswoche ohne Lärm (week without noise) from 6 to 12 March 1935.[4] The organizers of the Reichswoche ohne Lärm were especially concerned with noise at the workspace, which limited physical and mental work productivity, and argued that noise abatement had to be taken out of the private sphere and become a common law (Krug 1935).

The NS seizure of power led to a reorganization of scientific institutions. Jewish or politically unwanted scientists and other personnel were dismissed or sent into retirement and replaced by party members and other supporters of the new regime. The Führerprinzip (leader principle) gave supreme power to the institutes' leaders as structures of academic autonomy were abandoned. The reorganization had quite diverse implications for three institutions that I discuss in this chapter: the Physikalisch-Technische Reichsanstalt, the Institute for Sound and Heat Research, and the Heinrich Hertz Institute.

The NS appointed Johannes Stark in May 1933 as president of the Physikalisch-Technische Reichsanstalt, where he established acoustics as one of the main fields of activity (D. Hoffmann 1993; Kern 1994, 203–258). With Philipp Lenard, Stark was the most outspoken and influential representative of the anti-Semitic movement of Deutsche Physik, attacking quantum mechanics, the theory of relativity, and theoretical physics in general as un-German and Jewish (Beyerchen 1977, 103–167; Schröder 1993). Technical acoustics offered everything that Stark and other representatives of the

Deutsche Physik wanted "Aryan physics" to be in their imagination: experimental, applied, based entirely on classical physics, and useful for the German military.

Hermann Reiher became a member of the NSDAP in 1933 but was not an outspoken Nazi or representative of the Deutsche Physik like Stark. But Reiher could advance his career with the new regime. After Erich Regener, who opposed the NS and had a wife with Jewish ancestry, lost his power, Reiher finally succeeded in being appointed as a professor in technical physics at the Technische Hochschule Stuttgart in 1935. Regener was forced to leave his professorship in 1937. His opposition to Reiher's appointment was based on Reiher's scientific credentials and apparent lack of interest in scientific foundations.[5] The science policy of the new regime, however, did not ask for scientific foundations. The new leader principle also meant that Reiher did not require a scientific committee to appoint him. Reiher's credentials and his agenda of applied acoustics fitted rather nicely into the NS picture of how the discipline of physics should develop.

While Reiher was promoted, Karl Willy Wagner was sacked as director of the Heinrich Hertz Institute. The Ministry of Propaganda had taken over the Study Group for Oscillation Research, the patron of the Heinrich Hertz Institute, in 1934. Wagner had, like Reiher, joined the NSDAP in 1933. Like many of his colleagues, Wagner claimed after the war that he was deceived by Hitler's promise to unite the German people and did not take its anti-Semitism seriously.[6] While many scientists praised Wagner's achievements, others did not approve of his attitude. Jonathan Zenneck, for example, was a fierce opponent of Wagner, especially how Wagner monopolized the Heinrich Hertz Institute and dominated the field of oscillation studies in Germany.[7] Heinrich Fassbender, professor of electric oscillations and high-frequency technology at the Technische Hochschule Berlin-Charlottenburg since 1931, started a campaign against Wagner. Willi Willing, NS party official and professor of electro-economy from 1937, and Klaus Hubmann, director of the German Broadcasting Corporation, joined Fassbender in the campaign. In 1936, Wagner was dismissed as director of the institute and his right to teach at the Technische Hochschule revoked, all on grounds of alleged embezzlement of funds at the institute and for insubordination in the removal of Jewish employees. The *Akustische Zeitschrift* and the *Elektrische Nachrichten-Technik* were ordered not to print any of his articles.[8]

Wagner was not the only one removed from the institute. Hans Salinger, head of the Department of Telegraphy and Measurement, was dismissed for being a Jew and emigrated to the United States in 1936. Gustav Leithäuser, head of the Department of High-Frequency Technology, was downgraded in 1936 and had to leave in 1937 because, like Regener, his wife was of Jewish origin. After Willing and Fassbender took over the institute in 1936, the name "Heinrich Hertz" was removed. From now on it was simply called the Institut für Schwingungsforschung. This was only part of the NS campaign to dissociate from the legacy of Hertz, who was also partly of Jewish ancestry. The Verband Deutscher Elektrotechniker (VDE; Association of German Electrical Engineers), which had become part of the National Socialist Bund Deutscher Technik, lobbied to eradicate the name "Hertz" for the unit of frequency, which had been introduced by the German Committee at the International Electrotechnical Commission (IEC) only in 1930. The VDE tried to change it to "Helmholtz" to keep the initials "Hz."[9]

5.3 Volksempfänger and Gemeinschaftsempfang

Radio broadcasting and sound amplification at political rallies led politicians to address their audiences in new ways and created a new logic to frame political reality. Like musicians and film actors, politicians had to adapt to the new medium and learn to speak into the microphone. Political leaders had appeared on radio and used sound amplification in public addresses since the 1920s. They differed largely, however, in how they used the media and how they addressed their audiences. Franklin D. Roosevelt addressed the American people in carefully staged fireside chats. He aimed at transporting himself into the intimate atmosphere of American families (Wijfjes 2014, 163–165).

The intimacy of the living room was not exactly what Hitler or Goebbels wanted or in fact created. For the NS, radio allowed it to conflate sound, space, and time. Listening to the Führer was not supposed to be an individual experience but an organic synchronization of the masses. While Roosevelt created intimacy, Hitler created distance. Rather than bringing the speaker home, the NS propaganda speeches aimed to bring the listener out to become part of the spellbound masses. In fact, Hitler did not appear in the studio after 1933, and all his radio appearances were live broadcasts

(Wijfjes 2014, 166–170). For the NS propaganda apparatus, radio and sound amplification were interconnected technologies. The intimacy of the home was suspect to the NS ideologues, who much rather wanted their audience to attend communal radio broadcasts, in public spaces, schools, and factories.

NS propaganda identified radio broadcasting as the most important instrument of mass indoctrination and chose the Reichs-Rundfunk as one of the first institutions from which they removed everybody who was politically unwanted (Koch and Glaser 2005, 78–90). The Reichsministerium für Volksaufklärung und Propaganda (State Ministry of Public Enlightenment and Propaganda) realized that between 1933 and 1934, too many political speeches on the radio had led to a saturation for the listeners, and state-sanctioned music and entertainment gained a central role in radio programs until the outbreak of World War II (103–109). We cannot draw a sharp line, however, between propaganda and entertainment in the radio programs of the Third Reich, where indoctrination entered entertainment in a more subtle way.

The percentage of households with a radio receiver in 1933 was relatively low, at 25 percent. The Reichsministerium für Volksaufklärung und Propaganda initiated a campaign for the mass production and marketing of the Volksempfänger, a simple standardized radio receiver (König 2003a, 2003b). Gustav Leithäuser at the Heinrich Hertz Institute led the technical commission to oversee its development (König 2004, 36). The Volksempfänger, along with the Volkswagen (peoples' car), was perhaps the best known of a whole range of Volksprodukte (peoples' products), which were aimed at creating an NS consumer society.[10] By 1939 the number of radio receivers in German households had risen to 57 percent. But this was not such an overwhelming success since other European countries experienced similar growth rates of radio listeners despite the lack of elaborate marketing campaigns like that for the Volksempfänger. Moreover, radio coverage in Germany paled in comparison with that of the United States, with its staggering 1.5 radio receivers per household.[11]

To reach out to larger parts of the population, especially the working classes, the NS establishment developed concepts of Gemeinschaftsempfang (communal broadcasting) by setting up amplifier systems in factories, schools, and public offices. Whereas the Volksempfänger was a medium of propaganda and entertainment in the private family space (it could be turned off or tuned to different radio stations), communal broadcasting originated from the NS concept of Volksgemeinschaft, where listening to the speeches of Führer Adolf Hitler or other important figures was conceived as a collective experience (König 2003a, 98–100; Wigge 1934).

After Telefunken had taken over most of the electroacoustic activities from Siemens and AEG, it became the main supplier of electroacoustic amplification systems for the Gemeinschaftsempfang. Beyond communal broadcasting in factories and public buildings, the propaganda department of the NSDAP also developed plans to occupy open public spaces with *Beschallungsanlagen*, a system of loudspeakers on poles located at strategic sites in cities. Breslau, however, was the only city where a system was installed, one hundred poles equipped with speakers in 1938.[12]

5.4 Electroacoustic Amplification of Large Political Rallies

The NSDAP used sound amplification in their political rallies even before the seizure of power in 1933. In 1932 it rented Siemens and Halske loudspeaker vans to amplify speeches, songs, and party slogans (Birdsall 2012, 39). After January 1933, large loudspeaker systems became an important instrument to stage and orchestrate NS political rallies, such as the gatherings on the Tempelhofer Feld and the Nuremberg party congresses. The NS regime turned the 1936 Olympic Games, held in Berlin, into a media and propaganda spectacle of hitherto unknown dimensions.

Telefunken also supplied various Nazi organizations with loudspeaker systems for their mass gatherings, which became possible only with the electroacoustic amplification of speech. For the celebration on 1 May 1933 on the Tempelhofer Feld in Berlin, Telefunken supplied an amplification system of up to 4,000 watts of speech power, as reported in euphoric articles in the *Funk Bastler* and the *Telefunken-Zeitung*. According to the articles, more than a million people listened to the speeches of Hitler, Goebbels, and others.[13]

The amplification of sound at these very large gatherings in open spaces raised new challenges for acousticians. The installation of a single large loudspeaker next to the orator caused a delay in the sound reaching the audience. If commands were given over centralized loudspeakers, the resulting movement of the masses in the field, which could be up to a kilometer deep, would appear in waves.[14] This was not acceptable for the orchestration of rallies, which demanded perfect synchronization between the masses and their leadership. Sound was simply too slow. Perfect synchronization was achieved through a system of smaller speakers spread around the meeting place. This made sound travel faster than its speed and allowed the desired synchronization (Emde, Henrich, and Vierling 1937, 252–253). Telefunken developed a whole range of speakers for outdoor use. The most common speaker, however, was the round *Pilzlautsprecher* (mushroom speaker) mounted on poles (figure 5.2).[15] To test their loudspeakers, Telefunken operated a large open-air testing laboratory in Groß Ziethen, just outside Berlin (Schwandt 1934).

While the dispersed loudspeakers brought the sound more or less simultaneously to all participants of the mass gatherings, they caused other problems. The different loudspeakers interfered with each other and created echoes. To avoid interference, Telefunken developed the *Löschstrahler* (extinguishing loudspeaker). Two identical loudspeakers were mounted on one pole on top of each other (figure 5.3). These loudspeaker were switched in opposite phases and were supposed to extinguish each other's sound for

Figure 5.2
Telefunken Pilzlautsprecher (mushroom speaker) on the Wilhelmplatz in Berlin, 1936 (Photo: Stiftung Deutsches Technikmuseum Berlin, Historisches Archiv AEG FS 044–3–01, B 27068).

Figure 5.3
A Telefunken Löschstrahler (extinguishing loudspeaker) at the Reichssportfeld
stadium during the Olympic Games in 1936. The two loudspeakers were supposed
to extinguish each other at distant points for low frequencies to avoid disturbances
between different speaker poles (Photo: Stiftung Deutsches Technikmuseum Berlin,
Historisches Archiv AEG FS 044-3-01, B 26791).

low frequencies at distant points. The arrangement of extinguishing loudspeakers was used in the Reichssportfeld stadium during the Berlin Olympic
Games in 1936 (Meyer 1939, 70–71, fig. 47; Donhauser 2007, 161, 163).

The extinguishing speakers seem to have been used at the Reichssportfeld stadium only, indicating that their acoustic performance did not
meet expectations and that their development was abandoned. For the
amplification of music at the Olympic Games and other mass gatherings,
the dispersed loudspeaker systems were unsuitable. The generation and

amplification of music for the NS mass gatherings became a topic for Friedrich Trautwein and Oskar Vierling, two of the most prominent proponents of electroacoustic musical instruments. After the NSDAP seized power in 1933, electroacoustic music was treated with suspicion, and the development of electric instruments received little support. The Rundfunkversuchsstelle, where Trautwein had developed the trautonium, was starved of funding and finally closed down in 1935 (Donhauser 2007, 127–131). Both Trautwein and Vierling had become members of the NSDAP and had aligned their research and development of instruments and musical performance with the requirements and ideals of the new rulers. Trautwein now saw chamber music and string instruments as "characteristic for an epoch of a bourgeois music culture," an epoch that had passed (Trautwein 1937, 34–35). The NS mass gatherings required, according to Trautwein, a new form of music and music technology. Existing musical instruments were simply not able to supply the necessary sound volume. This volume could only be generated by an amplification system that amplified either music recorded by a microphone or electric musical instruments, for example, an electric organ (Trautwein 1937, 43).

The electric organ that Trautwein most likely had in mind was Vierling's organ, which gained prominence when it was played in the 1936 Olympic Games. The NS organization Kraft durch Freude supported Vierling's project, and the organ became known as the "Kraft durch Freude Grosstonorgel" (Vierling 1938a, 28, 17; Donhauser 2007, 148–151). The organ also became the subject of Vierling's habilitation thesis, published as *Eine neue elektrische Orgel* (A new electric organ) in 1938. Both Trautwein and Vierling argued that only large centralized loudspeakers were suitable for the amplification of music at large gatherings (Trautwein 1937, 42; Vierling 1938b). Vierling published about his experiences with a 5 kW amplification system employed during a midsummer night mass gathering of the NSDAP at the Imperial Castle of Nuremberg in 1938, where the loudspeakers were placed on the tower of the castle (figure 5.4). According to Vierling, this was the largest amplification system that had ever been set up in Europe (Vierling 1938b, 93).

In 1938 Vierling was appointed lecturer at the Technische Hochschule Hanover and left the Institut für Schwingungsforschung. By then, remilitarization and the requirements of war had completely taken over acoustics research and development for propaganda purposes and mass gatherings.

Figure 5.4
Loudspeakers mounted on the tower of the Castle of Nuremberg. From Oskar Vierling's report of a 5 kW amplification system used during a midsummer night mass
gathering of the NSDAP (Vierling 1938b, 94). The funnels of the loudspeakers can be
seen below the battlement of the tower.

He was soon involved in military research and got his own institute built
close to Ebermannstadt in Franconia in 1941, which was camouflaged as
a castle and named Burg Feuerstein. There, Vierling and his collaborators
worked on high-frequency technology, hydrophones, and speech encoding
(Donhauser 2007, 158, 233–234).

5.5   Remilitarization of Acoustics Research

While military research and development continued in the immediate
interwar period among the Allied countries, there was more of a holdup in

Germany.[16] Facing the seizure of the German fleet, the naval command had scuttled most of its ships in 1919. After the Treaty of Versailles, the German Navy was limited to ships smaller than 10,000 tons, and submarines were banned. The treaty prohibited the production of military aircraft and restricted the aircraft industry as a whole. While the restrictions of the Versailles Peace Treaty and the economic crises forced the industry to convert their production from military to civilian markets, limited amounts of military research and development continued during the Weimar Republic. An increasing remilitarization and renewed interest in acoustics research and development for military purposes had started before 1933.

Crisis and continuity of military research and development unfolded in different ways. As mentioned in chapter 4, delegates from the Reichswaffenamt or the Heereswaffenamt (Army Weapons Office) attended the annual seminars with professors from the Technische Hochschulen and universities at the Telegraphentechnisches Reichsamt and Reichspostzentralamt between 1926 and 1929.[17] The restrictions on U-boats and aircraft in Germany meant that German researchers had to reach out to international collaborations. Ludwig Prandtl worked with the United States National Advisory Committee on Aeronautics to continue research in Göttingen (Eckert 2006, 83–91). Research and development in underwater acoustics also relied partly on international collaboration and international markets. The loss of the military market landed the Signal Gesellschaft into financial difficulties. The Atlas Werke was more fortunate and could reinstate its collaboration with the Boston Submarine Signal Company.

The German Navy had founded the Nachrichtenmittel Versuchsanstalt (NVA; communication technology testing laboratory) in 1923, which kept contact with both Atlas and Elac (Rössler 2006, 26). While the merchant navy provided a market for echo sounding, Elac also sold listening apparatuses to the Spanish, Dutch, Swedish, and Japanese navies (30). Since the German Navy did not have its own submarines, the NVA carried out tests on submarines built for other navies (32–33).[18]

At the Siemens laboratories, military-related research continued during the Weimar Republic as well, as discussed in chapter 4. While Hans Riegger had turned his research in underwater acoustics to the development of microphones and loudspeakers for large amplification systems, Ernst Lübcke and Hans Gerdien continued to work on underwater sound and direction ranging, joined by Ferdinand Trendelenburg.[19] Lübcke,

Trendelenburg, and Gerdien also filed patents for Siemens for acoustic echo sounding and direction sensing for aviation.[20] While the Siemens researchers filed the acoustic patents related to aviation before the takeover by the NS in 1933, these patents were only issued and published thereafter.

Protagonists for Germany's remilitarization aimed at learning from the previous war. The early 1930s saw several publications of books that revisited the sound location experiences from World War I and that were directed toward a renewed military preparedness, anticipating a new war. Like the new interest in military research, these books started to appear prior to the Nazi seizure of power. In the autobiographic *Schallmesstrupp 51*, the NS propaganda author Martin Bochow reported from experiences in artillery ranging during World War I (Bochow 1933). NS party official and economic planner Heinrich Hunke's *Luftgefahr und Luftschutz* was, in contrast, a detailed study of the requirements of future aerial warfare and air defense, building on the experience of the previous war and analyzing the development of the field since (Hunke 1933). Erich Waetzmann published his *Schule des Horchens* in 1934 (Waetzmann 1934b). It was based on Waetzmann's detection and analysis of aircraft sounds with the unaided ear during World War I. As a pedagogical manual, it was meant to train and discipline a new generation of listeners. While its mission to create a new military preparedness is less explicit than in Hunke's *Luftgefahr und Luftschutz*, its relation to military attentiveness is difficult to miss (Encke 2006, 182–184).

Aviation research became a focus of NS science policy since it was central to the efforts to build up a strong air force. The Lilienthal-Gesellschaft für Luftfahrtforschung (Lilienthal Society for Aviation Research) and the Deutsche Akademie der Luftfahrtforschung (German Academy of Aviation Research) coordinated acoustics research, especially on silencing aircraft and on echo sounding.[21] During the 1930s, however, it became increasingly clear that radio detection and ranging (radar) was taking over the role of acoustic detection of aircraft. The rapidly increasing speed of aircraft made acoustic detection less and less relevant.[22]

While acoustic methods lost their importance for aircraft detection, they remained a priority for submarine warfare. When Johannes Stark took over as the president of the Physikalisch-Technische Reichsanstalt and established its acoustics laboratory in 1934, he had explicit military applications in mind. In a memorandum for the laboratory, he mentioned the development of apparatus for artillery ranging and aircraft detection, noise measurement of military vehicles, and communication systems for military pilots (D. Hoffmann 1993, 127; Kern 1994, 232). During the war, the acoustics laboratory worked almost exclusively on military projects. A water tank was built, and the laboratory measured the sound field of cruising ships to develop acoustic mines and acoustic guidance of torpedoes. In the course of the war, the workforce at the acoustic group doubled or tripled, finally reaching a size of sixty to seventy staff members (Kern 1994, 266–268).

In addition to the efforts in acoustics research at the Physikalisch-Technische Reichsanstalt, the Institut für Schwingungsforschung intensified its military collaboration under the leadership of Fassbender after its restructuring in 1936. The activity report of 1938 lists a group for underwater acoustics and mentions the construction of a water tank, without revealing any details of investigations undertaken.[23] After the German attack on Poland in September 1939, most of the institute's civilian research activities were stopped and redirected toward military-related research and development.[24] In 1941, a Vierjahresplan-Institut für Schwingungsforschung (Four-Year Plan Institute of Oscillation Research) dedicated entirely to military research and with a budget several times larger than its prewar levels supplemented the institute. Erwin Meyer directed the main activities of the Department of Acoustics, which included underwater acoustics research. The research and development at the Institut für Schwingungsforschung included the project with the code name Alberich, where U-boats were coated with a thin layer of rubber to absorb the asdic pulses, which were meant to reflect from the hull to enable detecting and tracking the U-boat. A similar project carried the code name Fafnir and used rib absorbers.[25]

Siemens also continued its activities in military research and development. Gustav Hertz had joined the Siemens Research Laboratories in 1935 after he was forced to leave his professorship at the Technische Hochschule Berlin because of his Jewish ancestry (Goetzeler 1994, 79–80). In 1940, Trendelenburg became the head of the Arbeitsgruppe Cornelius, a working group of the German Navy initiated by Ernst-August Cornelius to solve the "Torpedo Crisis" (the technical malfunctioning of the German torpedoes) and to develop new guidance systems and fuses for torpedoes (Trendelenburg 1966, 148–168). Karl Küpfmüller headed the development of an automatic guidance system for torpedoes. Gerdien was responsible for the development of electric fuses, while Gustav Hertz headed the development of acoustic fuses (Maier 2007, 2:702–710, 987–988, 1079).

We can conclude that after the NS gained power, there was a gradual shift in employing acoustics research and development, from orchestrating the National Socialist soundscape to remilitarizing, until the outbreak of World War II finally dominated acoustics research. To follow the efforts of acousticians during this war and the postwar period would be to open another book. In technical acoustics, as in other fields of science, several actors and their agendas received rather strong support from the NS establishment, while others were banned or persecuted. Most of the acousticians were not passive victims of the NS ideology and attitude toward science and technology but adapted their research interests and careers to the new regime.

6 Conclusion: The New Acoustics

6.1 Acoustics as Modern Physics

In the entire field of acoustics, extraordinary advancements have been achieved in the last years, where the impetus has come mainly through technical problems. We have here a prime example of how closely pure and applied physics are connected to each other today, and how strongly they influence and inspire each other. Pure acoustics has been woken out of its long sleep mainly through technical problems, and technical acoustics owes its upswing a great deal to the fact that it has put into service all achievements of pure physics.

—Erich Waetzmann, 1934[1]

What does the history of acoustics tell us about the discipline of physics and its transformations in the interwar period? What does it tell about its objects, its objectives, its practices, and its material and immaterial tools? As I argue in the introduction, the field of acoustics does not feature in our common understanding of "modern physics."[2] Technical acoustics and electroacoustics have been part of a tradition of applied, technical, and industrial physics that does not occupy a prominent place in existing histories of physics of the interwar period. This leaves us with a blind spot. While Johannes Stark mobilized technical acoustics for his alternative vision of what modern physics should be like, and tensions prevailed between communities of general and technical physics, most physicists of interwar Germany agreed to a picture similar to the one drawn by Waetzmann, which established a close relationship between the two. What I argue for is to place the new kind of technical acoustics and electroacoustics that emerged after World War I as integral to the narrative of the transformations that the discipline of physics underwent in the 1920s and 1930s.

Placing acoustics within our understanding of the dynamics of physics in the interwar period needs to be addressed on several levels. One level is the laboratory. During the 1920s and 1930s, corporate research laboratories became an important and attractive workplace for physicists. In Germany, more physicists worked in industry than in academia and public laboratories combined. Bernward Joerges and Terry Shinn have proposed the notion of research technologists in order to characterize certain types of relationships between scientists and industry. Research technologies such as electron microscopes, particle accelerators, and ultracentrifuges were developed in corporate laboratories or in proximity with industry (Joerges and Shinn 2000, 2–11). While acousticians in the interwar period generally do not correspond to the characterization of research technologists, technical acoustics and electroacoustics have been ahead of other fields in physics in terms of industry collaboration and the electrification of measurement in the interwar physics laboratory.

Electroacoustics, radio, and the amplification and processing of electric signals have effectively been coproduced or close to each other in the first decades of the twentieth century. By the early 1930s, electric amplification and signal processing became crucial to virtually all fields of experimental physics, as has been shown by the example of experimental nuclear physics in the early 1930s (Wittje 2003, 157–159). The field of electroacoustics developed between physics and electrical engineering. The same can be said about particle accelerator physics, where many of the accelerator builders, for example, Robert J. Van de Graaff, Rolf Widerøe, and John D. Cockcroft came from electrical engineering. John Heilbron and Robert Seidel have identified reliance on radio technology as the distinctive feature of the Berkeley Radiation Laboratory's approach to particle accelerators. Besides Ernest Lawrence, many of the laboratory's earliest workers had been radio hams. According to Heilbron and Seidel, "The Laboratory was so filled with radio waves that its members could light a standard electric bulb merely by touching it to any metallic surface in the building. Many cyclotron laboratories were to eke out their resources by cannibalizing old radio parts" (Heilbron and Seidel 1989, 127). Apart from the cyclotron accelerator laboratories, groups working with linear high-voltage accelerator designs relied largely on radio technology and practices from amateur radio. Merle Tuve's biographer, Thomas David Cornell, has stressed the importance of what he called "The Radio Pattern of Learning" in Tuve's professional development

(Cornell 1986, 13–26). Cornell describes Tuve's early years as a radio amateur with his boyhood friend Lawrence and emphasizes the importance of these early experiences with wireless for Tuve and Lawrence's later careers. Tuve had engaged in using radio waves to study the conducting layer of the atmosphere before he turned toward nuclear physics and accelerator building (151–162).

Jeff Hughes has disclosed the link between radio technology and the development of instrumentation in nuclear physics at the Cavendish Laboratory. He has argued against the narrative that experimental physics before World War II relied on sealing wax and string, still bearing the markings of small preindustrial science. The Cavendish Laboratory had a strong connection to the electrical industry, and modern and industrially made, highly elaborate products like radio valves had entered scientific instrumentation. Ernest Rutherford and James Chadwick commissioned radio amateurs and research students Eryl Wynn-Williams, Francis Alan Burnett Ward, and Wilfrid Bennett Lewis to construct an amplifier for Geiger-Müller counters. According to Jeff Hughes, for Rutherford and Chadwick, "familiarity with the latest developments in circuit and electronic hardware now [in the early 1930s] became one of the most valuable resources that a young physicist could bring to experimental nuclear physics" (J. Hughes 1998, 75).[3] In experimental acoustics, familiarity with electric circuits and amplifier hardware had already become an essential resource during World War I.

The industrial laboratory, electrical engineering, and radio were only one level of how technical acoustics and electroacoustics intertwined with broader developments in physics. Other connections operated on the levels of analogy, pedagogy, and mathematical representation of physical phenomena. The pedagogy of acoustics remained a pedagogy of mechanical vibrations within the basic curriculum of physics education. Understanding vibrations was utterly fundamental to all fields of physics, and acoustics offered a vivid and easily accessible approach to it. Outside this realm of physics pedagogy, sound became increasingly electrified. Inherently acoustical concepts of noise entered the theory and practice of electrical systems and communication engineering. From there, notions of noise spread into all fields of science.

The analogy between acoustic and electromagnetic oscillations was explicitly productive in the transfer of mathematical formalism and methods since wave equations and wave propagation are at the heart of most fields of physics. Rayleigh's *Theory of Sound* became very useful outside the field of acoustics in the development and interpretation of Maxwellian electrodynamics, as I discuss in chapter 2. Field theory radiated back from electromagnetism to acoustics during World War I, when Walter Hahnemann and Heinrich Hecht introduced the sound field in an analogy to the electromagnetic field. The wave formalism of acoustics as developed by Rayleigh and others remained useful in similar ways for quantum mechanics and the theory of relativity in the twentieth century. When discussing gravitational waves in his *Mathematical Theory of Relativity*, Arthur Eddington could refer to Rayleigh's "well-known solution" of a given wave equation (Eddington 1923, 129; Warwick 2003, 484n114).

Erwin Schrödinger's formulation of quantum mechanics as wave mechanics made the transfer of wave formalism from acoustics specifically attractive.[4] Johan Holtsmark and Hilding Faxén derived a quantum mechanical theory for the Ramsauer-Townsend effect, the scattering of slow electrons by atoms in noble gases, during a stay at the Bohr Institute in Copenhagen in 1927. They applied Schrödinger wave mechanics and analog formalism from acoustic wave scattering, taken again from Rayleigh's *Theory of Sound* (Faxén and Holtsmark 1927; Wittje 2003, 145). The analogy with acoustic wave theory was also used to illustrate elsewise abstract concepts of quantum mechanics. Alfred Landé, a former student of Arnold Sommerfeld who had worked on spectroscopy and atomic theory, presented a series of lectures on the new wave mechanics during a stay at Ohio State University in 1929. When Landé introduced Heisenberg's uncertainty principle, he compared the uncertainty of time and energy to the acoustic diffraction limit, determining the minimum time span necessary to identify a particular frequency with certainty.[5]

Methods and concepts of wave mechanics could move both ways, from acoustics to quantum physics, but also back to acoustics. George W. Stewart at Iowa State University picked up Landé's idea and proposed an equivalent uncertainty principle in acoustics in 1931.[6] To be sure, Stewart did not suggest that physical acoustics was based on the foundations of quantum mechanics. He pointed out that the uncertainty principle in acoustics was entirely different from the origin of the Heisenberg principle. The uncertainty principle in acoustics related to the time span the human ear needed to detect a change of frequency in time. The minimum time span required to detect a frequency variation depended directly on the frequency, with lower frequencies requiring more time. Stewart pointed to experiments carried out by Vern Knudsen and Harvey Fletcher, implying that the human ear detected frequency variations close to the theoretically possible limit but failed to do so when it came to sound intensity variations. Fifteen years later, Dennis Gabor drew on Stewart's proposal in his "Theory of Communication" of 1946 and in a paper entitled "Acoustical Quanta and the Theory of Hearing" in the journal *Nature* in 1947 (Gabor 1946, 1947). While Stewart's paper did not seem to have attracted much attention, Gabor's publications became a landmark in the theory of human hearing and communication, which developed after World War II.

Unlike Landé, Gabor did not come from the schools of quantum mechanics but came from industrial research. He had studied physics and electrical engineering in Budapest and at the Technische Hochschule Berlin before he started to work at the physical laboratory of Siemens and Halske in 1927. As a Hungarian Jew, Gabor left Germany after 1933 and moved to Britain to work in the laboratories of British Thomson-Houston (BTH). At Siemens and BTH, Gabor worked on gas discharge, plasma physics, and electron microscopy. As Friedrich Hagemeyer has argued, Gabor's theory of communication was part of the overarching research program of oscillation research that prevailed in Germany (Hagemeyer 1979, 40–41). Gabor did not carry out any prominent work in acoustics at Siemens or at BTH. His publications of 1946 and 1947 nevertheless exhibit his electroanalog line of argument and his familiarity with the research on electroacoustic media technologies in corporate laboratories. In his 1947 paper, Gabor cited the experiments of electroacousticians from Bell Laboratories and a group at Telefunken Laboratories (Shower and Biddulph 1931; Bürck, Kotowski, and Lichte 1935a, 1935b). Gabor's theory of communication grew out of the efforts to optimize electroanalog speech transmission on telephone lines, and economic methods of frequency conversion and reproduction of sound and speech in sound motion picture technology and other analog systems of sound storage and transmission (Gabor 1946, 445–446).

Gabor's approach of acoustical quanta was based on Stewart's uncertainty principle in acoustics, taking into account the duality of the time and frequency pattern of an acoustic signal (1946, 432). The theory of hearing, which was connected to the names of Ohm and Helmholtz, Gabor argued, assumed that the ear would perform a Fourier analysis, selecting a sound signal into its spectral components. But this frequency analysis ignored the time pattern of the acoustic signal. In 1924, both Harry Nyquist at Bell Laboratories and Karl Küpfmüller at Siemens central laboratory had determined the relationship between the maximum number of telegraph signals that could be transmitted over a line and the waveband width (Gabor 1946, 429–430). In the first part of his 1946 paper, Gabor discussed the transmission of elementary signals or pulses. In the second part of the paper, he came to the questions of signal transmission and human hearing. What were the limits of signal transmission for intelligible speech? And since human hearing was masking rapid frequency variations, what were the limits of transmission for the reproduction of human speech and music that the ear could not distinguish from the original? (442)

Both Stewart and Gabor stressed that there was no uncertainty principle in physical acoustics as there was in quantum mechanics. But there was uncertainty in human hearing. What Stewart and Gabor described was not part of physical but of "subjective acoustics" (Gabor 1947, 591). Gabor suggested that it would be interesting nonetheless to both physicists and physiologists. While Stewart had insufficient data to determine whether human hearing was guided by the acoustic uncertainty principle, Gabor discussed experiments on the variation of pitch by Edmund G. Shower and Rulon Biddulph at Bell Laboratories in 1931, as well as by Werner Bürck, Paul Kotowski, and Hugo Lichte at the Telefunken Laboratories in 1935. Gabor concluded that the ear had the capacity to adjust to "the finest details of the sound patterns offered to it" (Gabor 1947, 593). Some of the experimental results could be explained "mechanically" by masking, resulting from the damping of the ear resonators, while the second mechanism was not mechanical and was probably located in the brain. Gabor suggested that the first "mechanical" mechanism was sufficient to explain the intelligibility of speech, but that it was the second, "non-mechanical" mechanism that made us appreciate music.

Not everybody felt at ease with the concepts of quantum mechanics. I have suggested that technical acoustics and electroacoustics served as a model subfield of technical physics, which at least for some physicists offered a different vision of what modern physics could be like: not abstract and theoretical but applied and technical. Quantum mechanics and the theory of relativity, usually associated with modern twentieth-century physics, remained distant, abstract, and *unanschaulich* (nonintuitive) for the general public as well as for many physicists. But everybody could experience the transformations of acoustics. Electro-acoustic technologies, along with an increasing acoustic manipulation of architectural space, changed the soundscape of urban spaces and brought reproduced and amplified sound into public and private environments. In his forty-four-page booklet *Vier Fragen an den Weltäther* (Four questions posed at the world ether) of 1954, Heinrich Hecht actually proposed to base the foundations of physics on technical physics and Schwingungs-forschung rather than on the abstract concepts of quantum mechanics and relativity. Hecht mobilized analogies between the propagation of sound and light when he presented his explanation for the independence of the speed of light from the velocity of the light source (Hecht 1954, 25–44). Hecht, who was seventy-four years old at the time of the booklet's publication, had clearly moved himself far from what could be seen as acceptable physics by any measure.[7]

By and large, physicists argued, like Waetzmann, for a strong relation-ship between technical physics—or technical acoustics, for that matter—and "basic" and theoretical physics rather than for a separation between them. Where physicists tended to disagree was on the nature of this relationship. Waetzmann described it as mutual inspiration (Waetzmann 1934a, 1:v; Meyer and Waetzmann 1936, 114). Carl Ramsauer presented a different picture in his article "Die Schlüsselstellung der Physik für Natur-wissenschaft, Technik und Rüstung" (The key position of physics for natu-ral science, technology, and armament) in *Die Naturwissenschaften* in 1943. In the picture drawn by Ramsauer, the flow of inspiration and knowledge was pointed clearly only in one direction, from "Grundlagenforschung" (basic research) in the core, through "allgemeine Physik" (general physics), to the other peripheral sciences and technical disciplines (figure 6.1). In Ramsauer's image, electroacoustics was one of the fields of technology where the flow pointed from general physics. In the diagram, electroacous-tics represented a field of physics that was in the process of breaking away and becoming its own technical discipline.

The article of 1943 was part of Ramsauer's agenda as chairman of the German Physical Society to rehabilitate theoretical physics in Germany within the National Socialist establishment after the attacks of the repre-sentatives of the Deutsche Physik, and to strengthen the position of basic physics within German wartime research and development. Ramsauer,

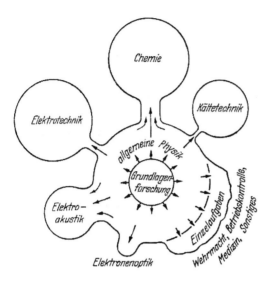

Figure 6.1
Carl Ramsauer's diagram of the relations of "basic research" and "general physics"
with technology, chemistry, military, and so forth (Ramsauer 1943, 286).

however, republished an edited version of the article in 1949 in which he
removed every trace of the military aspect but kept the hierarchy of the
science-technology relationship.[8] The 1949 version showed the same diagram without the reference to the "Wehrmacht" (the German armed forces
during the Third Reich).[9] While many scientists shared Ramsauer's views
about the relationship between physics and technology, which also placed
chemistry as a science that grew out of and required a supply of knowledge
from general physics, these positions were controversial within the community of physicists.[10]

6.2   Topographies of Science and Discipline Building

In the 1943 article and in its republished version of 1949, Ramsauer presented electroacoustics as an example of a scientific field that was in the
process of breaking away from physics and becoming an independent technological discipline (Ramsauer 1943, 286). Did electroacoustics eventually
separate itself from physics and electrical engineering and become its own
independent discipline? The interwar period saw the emergence of a community of acousticians that defined and redefined the field of acoustics and

its boundaries. The community established its own scientific journals, its handbooks and textbooks. Despite Ramsauer's predictions, however, electroacoustics did not become its own discipline as anticipated. Though the Acoustical Society of America was founded in 1929, the German Acoustical Society only came into existence in 1988. The *Akustische Zeitschrift* was discontinued in 1945. Acoustics remained an active research discipline within German communities of physics and electrical engineering after World War II and throughout the second half of the twentieth century. The place of acoustics within the physics community was reinforced with the appointment of Erwin Meyer as chair of the new Drittes Physikalisches Institut (Third Institute of Physics) at the University of Göttingen in 1947.[11] Here he started a school of acoustics research that is still active, both in Göttingen and in other places in Germany.[12] The Drittes Physikalisches Institut continues to identify with the legacy of Erwin Meyer and the program of oscillation research.[13] The Institute of Acoustics and Speech Communication at the Technical University of Dresden, the successor of the Technische Hochschule, identifies with the legacy of Heinrich Barkhausen in similar ways (Költzsch 2003). With the division of Berlin after World War II, it lost most of its electrical industry and its status as the German center for acoustics research. The Heinrich-Hertz-Institut für Schwingungsforschung was continued in West Berlin as well as in East Berlin, with both institutions bearing the same name. The reunified Heinrich-Hertz-Institut für Nachrichtentechnik has become part of the Fraunhofer-Gesellschaft.

Scientific communities, scientific institutions, and the production and consumption of scientific knowledge and technologies operate on local, regional, national, and trans- or international levels. Instead of attempting to understand developments of science and technology, such as technical acoustics, in an apparently fixed national context, a more fruitful approach is to try to understand how these developments of science and technology and the establishment of modern nation states and national identities have interacted.[14] Radio broadcasting and sound motion pictures acted as agents of nation building in this sense. Members of the community of technical physics aimed at mobilizing the discipline of physics in a certain way to rebuild Germany's industry and strengthen its international competitiveness after, according to these physicists, a disastrous war had shaken Germany's geographical, industrial, and political landscape. The German nation state had to be reinvented after 1918, and scientific and technical development was an actor rather than a passive object in this process, leaving neither science nor the state the same. Similar conclusions can be drawn for the takeover of Germany by the National Socialist dictatorship in 1933 and its redefinition of the nation.

By the 1930s a large international community working on the new acoustics research included members not only from all over Europe and North America, but also from India and Japan. Acoustics was both local and global. There was a local demand for acoustic expertise in most parts of the world, partly because of the local implementation of new electroacoustic technologies, like radio broadcasting and sound motion pictures, but also because of technologies that need silencing, like automobiles and airplanes. At the same time, acoustic knowledge traveled fast around the world through sound technologies, publications, and visits of scientists and engineers. On the one hand, some of the acoustic knowledge was specifically local or regional because of, for example, local or regional construction materials and methods, local languages and listening habits, or local musical cultures. On the other hand, building materials, listening habits, and musical cultures became more uniform, partly through the spread of acoustic practices and technologies. Local or national interests also had to be defended. Industries wanted to keep or expand the markets of their products, as we have seen from the example of sound motion pictures. The German electrical industry had a large international market and was exposed to tough international competition not only from the American electrical industry but also from other European actors, such as the Dutch Philips company.[15] The development and eventual success of sound motion pictures in Germany cannot be understood without comprehending the American development and its effect on the German sound motion picture market (Jossé 1984; Mühl-Benninghaus 1999).

National regimes played an important role when it came to radio broadcasting, military interests, and the takeover of the National Socialist Party in Germany in 1933. While it makes sense to place the history of acoustics in a national frame in certain analyses, this frame falls short in other cases. The German speaking and publishing community of acousticians, for example, extended beyond the German borders. The institution of the Technische Hochschule was not limited only to the German states but extended to other countries, like Austria, Switzerland, the Netherlands, as well as Eastern European and Nordic countries.[16] With the concept of a Technische Hochschule as a university of technical sciences, physicists in other countries established technical physics as a distinguished career and professional identity at their Technische Hochschulen in the interwar period.[17] It is therefore not surprising to find physicists at the Technische Hochschulen outside Germany being part of the agenda of technical acoustics, such as Franz Aigner at the Technische Hochschule in Vienna; Franz Max Osswald at the Eidgenössische Technische Hochschule in Zurich; Adriaan Fokker at the Technische Hogeschool van Delft, the Netherlands; and Johan Peter Holtsmark at the Norges Tekniske Høgskole in Trondheim, Norway.[18]

The Technische Hochschule was only one type of institution that defined technical acoustics. If we compare the research dynamics and processes of knowledge production in the field of acoustics among the different types of institutions in Germany during the interwar period, we can identify noticeable differences between corporate laboratories of the electrical industry (like the AEG and Siemens research laboratories) and public academic institutions (like the Heinrich Hertz Institute or the Schwachstrom Laboratory at the Technische Hochschule in Dresden). In the electrical industry, acoustics research was closely related to the development of electroacoustic devices, such as loudspeakers, microphones, and magnetic tape recorders, as well as technological systems such as telephony and sound motion pictures. The complexity of these technologies and systems brought acousticians together with communities and practices from other disciplines. Beyond addressing scientific and technical aspects, researchers in the corporate laboratories were confronted with issues of product design, manufacturing, cost efficiency, and user friendliness. In the case of the AEG Forschungsinstitut, the initial production line of sound motion picture cameras was even moved into the research laboratory. Many of these rather broad technical issues made it into the reference literature of technical physics and technical acoustics.[19] The Heinrich Hertz Institute, in contrast, though an applied research institute with close ties to industry, could develop Schwingungsforschung under a different kind of transdisciplinary agenda that did not have to be tied to any one technological system or production line, while departments at universities and Technische Hochschulen remained within their disciplinary agenda.

6.3   Sound Measurement

What did it mean to be an acoustician in the first decades of the twentieth century? The institutional setting for acoustics research had changed from the university setting of Helmholtz and the instrument workshop of Koenig in the nineteenth century to the Technische Hochschulen, public research laboratories, and the electrical industry after World War I. Acoustics had been transformed on an institutional level but arguably even more so on the level of practices. The kinds of sounds that were measured as well as the practices and instruments of sound measurement had changed substantially by the 1920s. What kinds of skills did acousticians need to be able to produce, control, and measure sound?

The most common measurements that acousticians carried out can be characterized as measurement of pitch or timbre, of sound intensity, and of time. Time measurement concerned both the time span between separate sound events and the time span for which one sound event occurred. In the course of the nineteenth century, the acoustic properties of musical instruments and music performance, especially pitch and beat, were increasingly standardized and quantified through the collaboration of scientists and musical instrument makers (Jackson 2006). Musical listening, musical literacy, and musical practice were essential skills to perform acoustic measurements in the times of Helmholtz and Rayleigh. Most of the scientific instruments of acoustics were closely related to musical instruments and to the practice of music as well. Musical listening was regarded as absolute listening. Helmholtz, for example, did not question the precision and impartiality of his own listening but turned it into the reference for his scientific enterprise.

In the 1920s and 1930s, acousticians no longer believed in absolute listening or the impartiality of their own listening. Human observers were unable to produce precise measurements of pitch, intensity, or time since the results varied from observer to observer, and even for the same observer from day to day. Musical listening was not required anymore; musical sounds had ceased to be the main reference for acoustics research. Instead, acousticians needed skills in electrical engineering and especially radio. They needed to understand transducers, electric circuits, amplifiers, circuit design, and electric filters in order to control and manipulate sound and sound measurement. The electrification of sound measurement was directly related to the transition from subjective to objective measurement during World War I. The Bull sound-ranging apparatus for artillery ranging with the Tucker hot-wire microphone was arguably the first successful system of objective sound measurement. The Tucker hot-wire microphone also exemplifies the meaning of "objective sound measurement." It did not indiscriminately record "all sounds that were out there," as one might think. The Tucker microphone was explicitly selective and recorded only the very specific low pitch sounds of the firing of heavy artillery.

One might argue that the Tucker microphone was a resonant microphone, or a signal receiver in the language of German hydrophones during World War I, while nonresonant microphones, or noise receivers, recorded sound more indiscriminately. But nonresonant hydrophones were also not supposed to record "all sounds that were out there," especially not the splashing sounds from the towing of the microphones or the noises of the observing ship. It was the acoustician's subjective goals and the characteristics of the microphone design that decided which sounds were recorded and how these were recorded. Objective sound measurement meant nothing more or less than mechanical, electrical, or optical machines recording pitch, intensity, or time instead of human observers. Carl Stumpf and Erich Moritz von Hornbostel had claimed objective sound measurement for their phonograph recordings in comparative musicology before World War I. But the phonograph recordings still required human listeners to analyze them. As Barkhausen argued in 1911, human skills of listening were essential to interpret the distorted sound reproductions of the phonograph.

Electroacoustic devices revolutionized sound measurement and its technology. Tuning forks, organ pipes, and resonators had to make space for radio and telephone technology.[20] But though the human listener could not be trusted to make precise measurements, it took until around 1930 before microphones combined with amplifiers could outpace the human ear in measuring sounds of low intensity. The Bull sound-ranging apparatus did not measure sound intensity but measured the time span between separate sound events of high intensity. Erwin Meyer pointed out that until electric amplification was applied, there was no chance to measure sound adequately by mechanical means, since "the power expended in conversational speech of an average speaker is of the order of 10^{-5} watt, which means that a speaker would have to speak uninterruptedly for 150 years in order to expend sufficient energy to boil the water to make one cup of tea" (Meyer 1939, 2).

While electroacoustic measurement of sound intensity remained inadequate or impossible for sounds of low intensity until the late 1920s, Max Wien had used the telephone as a sound source for loudness measurements already in the 1880s. Both Rayleigh and Max Wien observed that the relationship between the physical sound intensity and the subjectively observed loudness of a sound was not straightforward. Emily Mary Smith and Frederic Charles Bartlett used buzzer circuits to produce sounds to train hydrophone operators during World War I. Harvey Fletcher at Bell Laboratories and Heinrich Barkhausen at Technische Hochschule Dresden used electroacoustically generated sounds in their investigations into the relation between the physical intensity of a sound and its perceived subjective loudness.[21] Fletcher was not interested in the listening skills of trained musical listeners but wanted to know how "people with normal listening" listened. People with normal listening were, after all, the customers of Bell Telephone.

Of course, the sounds that Helmholtz observed in the *Sensations of Tone* in 1863 were very different from the sounds that acousticians dealt with during and after World War I. Helmholtz observed only musical sounds, which he defined in opposition to noise. Most of the sounds that scientists dealt with in their measurements during and after the war were neither musical nor harmonic. The kinds of measurement apparatuses that scientists built and used were directly related to the sounds they dealt with. Smith and Bartlett's buzzer circuit to train submarine hydrophone operators and Barkhausen's meter to measure sound intensity created not music but noise.

6.4 Concepts of Noise and Their Diffusion

As I argue in chapter 1, Hermann Helmholtz and Lord Rayleigh were concerned only with musical sounds, which they regarded as harmonic. They defined noise as nonharmonic, nonmusical sounds. This definition of noise as the opposite of musical sounds became problematic even before World War I. During the war, scientists turned from musical sounds to the industrial sounds of the battlefield. The pair of categories "music and noise" was replaced by the categories "signal and noise." Noise abatement became important to acousticians.

The rise of electric communication technologies before and during World War I brought a whole new category of noises that originated in electric systems and could be heard in the telephone and in the headphones of a telegraph or wireless operator. The sound of a singing or speaking arc produced a constant hiss that could be reduced but not eliminated. But the device that truly revolutionized electroacoustic communication was the amplifier tube, invented around 1906 for the amplification of telephone signals. Schottky identified the lower limit of the electrothermal distortion in amplifier tubes in 1918, describing the sound it created in the telephone (Schottky 1918). Telephone amplifiers were the electric systems Schottky analyzed for Siemens in his daily practice. This made the transfer of acoustical language to amplifiers and their circuits a natural one. While Carl A. Hartmann, who worked with Schottky at Siemens, described the sound as the Schroteffektton (shot effect tone) in 1922 (Hartmann 1922, 71), John Bertrand Johnson of Bell Laboratories in 1925 finally described it as a noise (J. B. Johnson 1925, 71–72). By the mid-1920s, the notion of noise had become a characteristic to describe random distortions in electric systems. Thermal noise generated by amplifier tubes was by far not the only or the defining aspect of "electric noise," or Rauschen in German.[22] Carbon microphones, which were the most common microphones in the 1920s, created a characteristic noise (*Mikrophonrauschen*) caused by the carbon contacts (Trendelenburg 1927, 557).

Harvey Fletcher became one of the most influential acousticians in the United States. His widely cited monograph *Speech and Hearing* was published in 1929, after he had organized the inaugural meeting of the Acoustical Society of America at the Bell Laboratories headquarters in New York in December 1928 (E. Thompson 2002, 105–106). In *Speech and Hearing*, Fletcher divided the world of sound into three categories: speech, musical tones, and noises, stating that "[p]ractically all types of sound which cannot be classified as speech or musical tones come under this classification [noise]" (Fletcher 1929, 99). His definition of noise remained vague: "Those sounds to which no definite pitch can be assigned are usually classified as 'noise'" (Fletcher 1929, 99), but then further down on the same page, he explained, "When transmitting speech or music either directly to an audience in a large hall or over an electrical system, such as a radio or a telephone system, there is always an interference to the proper reception of such speech and music, due to other sounds being present. These extraneous sounds which serve only to interfere with the proper reception are designated by engineers as 'noise.' With such a designation, the sound may be either periodic or non-periodic as long as it is something that would be better eliminated."

Fletcher did not dissolve or even discuss the obvious discrepancies, if not contradictions, in those very different definitions of "noise" but went on about the measurement of noise, both in its spectrum and its intensity. He presented the Western Electric audiometer through which, similar to the Barkhausen noise meter, "an individual with normal hearing" measured the sound by comparing it to a given sound reference (Fletcher 1929, 104). A number of different versions of the Western Electric audiometer were used in the investigations of the Noise Abatement Commission appointed by the New York City Health Commissioner in 1929 (E. Thompson 2002, 161–164).

Noise abatement initiatives and societies had emerged already around the fin de siècle in North American and European cities. Their founders and members came from urban bourgeois society and often had academic backgrounds. The noise abatement initiatives were closely related to the hygienists' movement and argued against the rise of industrial noises as well as the perceived decline of decent behavior (Braun 1998; Bijsterveld 2008, 54–136). Acousticians, however, did not pay attention to noise in the city before the late 1920s. In the UK, the Science Museum at South Kensington hosted a Noise Abatement Exhibition in 1936. The Research and Development Committee for the exhibition was chaired by George William Clarkson Kaye, superintendent of the Physics Department of the National Physical Laboratory (Science Museum 1935). Other major contributors to the exhibition included the Industrial Health Research Board and with it Bartlett, who had written on *The Problem of Noise* in 1934. There he had posed the question "What is Noise?":

Noise is any sound which is treated as a nuisance. This neither appears nor is a very exact definition; but, from the point of view of the present discussion, it is probably as good a one as can be obtained. Certainly the physical definition of noise as sound resulting from stimuli which cannot be resolved into periodic vibrations is, apart from other difficulties, a hopeless one. At the basement of the laboratory in which I work, an electrical generating plant is in constant use. The dynamos produce a musical tone which can be heard all over the building. I have yet to meet any research worker in the laboratory who fails to treat this musical tone as a noise. Even a tuning-fork, yielding only the purest tone, can be a horrid nuisance on occasion. Let anyone sit in a sound-proof room—as I have repeatedly done—for two hours at a stretch and listen to the sound of an electrically maintained tuning-fork coming intermittently throughout the whole of that period. At the end of the experiment he may feel inclined to use much the same language of the tuning-fork as that which the less restrained members of the community employ when they write to newspapers about clattering milk pans, clanging bells, or shattering motor-horns. (Bartlett 1934, 2–3)

In his discussion of the nature of sound, Bartlett brought together the two concepts of noise, one "noise as a non-periodic vibration," the other "noise as a nuisance," which in the German language are separated by the notions of Geräusch and Lärm. The engineer Richard Berger made this point clear in his chapter on noise abatement in the *Handbuch der Experimentalphysik* of 1934: "Noise [in the sense of nuisance; Lärm] is every unwanted audible sound, indifferent of whether it is a tone or a noise [in the sense of a nonperiodic sound; Geräusch]. The notions of noise [in the sense of nuisance] and noise [in the sense of a nonperiodic sound] are not identical."[23]

In Germany as well, antinoise societies emerged around the turn of the century, such as the Verein gegen Straßenlärm (Society against Street Noise) in Nuremberg in 1898. But scientific societies, as Berger pointed out, emerged only after World War I.[24] Following the example of the New York Noise Abatement Commission, the Verein Deutscher Ingenieure (VDI; Association of German Engineers) launched a Committee on Noise Abatement (Lärmminderungsausschuß) in 1931 and appointed Karl Willy Wagner, the director of the Heinrich Hertz Institute for Oscillation Research, as its chair (Wagner 1931). In 1933, the German Commission for Units and Formula (Ausschuß für Einheiten und Formelgrößen; AEF) drafted instructions on how to demarcate and define the two different notions of noise, as Wagner explained:

"Noise" [in the sense of nuisance; Lärm] is often equaled with "noise" [in the sense of a nonperiodic sound; Geräusch]. But these are two different things. The [German] Commission for Units and Formula defines in the following way (Draft no. 37) [*Elektrotechnische Zeitschrift 54* (1933): 783]:

"Noise" [in the sense of a nonperiodic sound; Geräusch]: Sound vibrations which are composed of a continuous spectrum of tones, or of a spectrum with very many simple tones of arbitrary height.

"Noise" [in the sense of nuisance; Lärm]: Any type of sound vibration that distorts a deliberate sound recording or silence.

"Noise" [in the sense of a nonperiodic sound] is, according to this definition, a purely physical value; the measurement of Noises [in the sense of nonperiodic sounds] can therefore, at least in principle, not lead to any difficulties. The concept of Noise as Nuisance, in contrast, contains next to physical also psychological elements; these cannot be determined by physical measurement, or at the best determined indirectly. (Wagner 1936a, 12)[25]

With these definitions, the AEF had demarcated two very different definitions of noise. One of them, Noise as Nuisance, was identical with Bartlett's definition and included the different types of noises that the noise abatement movement was dealing with. Wagner divided the nuisances into the categories of "street noise," "business and workshop noise," and "domestic noise" (Wagner 1931, 161). The category of domestic noise included musical instruments, loudspeakers, record players, and dancing music (158). Noise as Nuisance was, by definition, a subjective category. The second definition, Noise as Nonperiodic Sound, was, by contrast, defined as a purely physical value and therefore an objective category.

The "purely physical value" of noise as nonperiodic sound was, however, not as unproblematic as the clear-cut definition of AEF might suggest. Acousticians were still convinced that the very existence of a subdiscipline of physical acoustics was justified by its importance for the sensation of human hearing, and therefore for human culture. Ferdinand Trendelenburg, physicist and chief acoustician of the Siemens and Halske laboratories, stated in 1935 that it would be possible to have a pure "physics of sound," but that it would be scientifically unsatisfying (Trendelenburg 1935, 1). He added:

The compound tone of a piano, for example, is not strictly periodic, in the moment of strike [of the piano string], the compound tone is mixed with the noise (Geräusch) of the hammer …; nevertheless, in linguistic usage we would always call the sound of the piano a compound tone [and not a noise], for the ear in this case the tone-like characteristic is the preeminent characteristic [of the sound of the piano]. (Trendelenburg 1935, 19–20)[26]

With this statement, industrial physicist Trendelenburg takes us straight back to Helmholtz and his preferred instrument, the piano (Hui 2013, 59). Much had changed since the times of Helmholtz, and technical problems had come in the forefront of acoustical research. But whether the sounds came from the piano, the radio, or the telephone systems of Siemens and Bell, there was always someone who listened. A "pure physics of sound" where nobody was listening was deprived of its scientific substance, both for Helmholtz and for Trendelenburg.

The industrialization of acoustics had lasting consequences for the importance as well as the meaning of noise. From being defined as the antipode of what acousticians were studying, it became a focus of sound studies during World War I and thereafter. From our histories, we can make out essentially four different concepts and definitions of noise:

1. Noise as nuisance. In this definition noise can be periodic or nonperiodic. This was the definition given by Bartlett in 1934 and the definition of Lärm by the German Commission of Units and Formula in 1933.

2. Noise as a sound other than music or speech. Both Helmholtz and Rayleigh used the notion of noises as sound other than musical sounds. Fletcher extended this to noises as sounds other than speech and music. While these definitions seem similar, their foundations were utterly different. For Helmholtz, sound was music. For Fletcher, speech and music had to be transmitted through telephone or broadcast by radio. Everything but speech and music was a distortion. For Helmholtz and his contemporaries, listening was an individual experience that gained authority through the musically trained ear of the acoustician. Fletcher and his contemporaries, in contrast, measured the sound perception of the average ear, which was gained by taking the arithmetic mean of a large number of test subjects (Fletcher 1923, 302).

3. Noise as the antipode of a defined signal. In this conception, noise can be perceived as nuisance but also as carrying important information, depending on the observer's attention. This notion of noise came to the fore in the sound-signaling and sound-ranging operations of World War I as described by Aigner and others. This notion of noise, which at times has to be suppressed for the measurement of the signal and at times has to be listened to, can be found in measurement theory and measurement procedures of contemporary science.

4. Noise as a vibration, either electrical or mechanical, which cannot be dissolved into periodic harmonic vibrations. This is a common definition of noise as Geräusch, given by Helmholtz in 1863 as well as by the German Commission of Units and Formula seventy years later. This notion of noise as nonharmonic seems to satisfy physicists and engineers because of its apparently objective character. A complete reduction of the science of acoustics to its "objective physical core," however,

did not make sense for acousticians like Helmholtz and Trendelenburg, as I have argued. For them, physical acoustics was necessarily intertwined with the essentially subjective human perception of sound, which depended not only on physiology but also on cultural and aesthetic aspects and the changing soundscape of the modern world.

6.5   Electroacoustics as a New Way of Thinking and Talking about Sound

The transfer of concepts of noise from acoustics to electrical systems was part of electroanalog thinking, which transferred and translated all kinds of entities from acoustics to electromagnetism and vice versa. Acoustic systems could be translated into electrical systems, represented by equivalent circuit diagrams, and understood and manipulated by circuit analysis.

William Henry Eccles, who was known for his work on electromagnetic wave propagation and circuit design, gave a presidential address to the Physical Society of London on 22 March 1929. In his address he brought the novelty and importance of the "new acoustics" that had arisen in the previous decade to the attention of British physicists.[27] In addition to the prominence that electroacoustic measurement technology had gained in acoustics research and the rise of electroacoustic media technology, Eccles pointed out the conceptual aspects that had transformed electroacoustics into a new language of sound:

Besides these tangible adjuncts to the technique of experimental acoustics, electrical science has brought subtle assistance to the more theoretical aspects of the subject. This comes about because vibration phenomena of all kinds approximately satisfy the same linear differential equations. Inasmuch as the study of electrical vibrations in well-defined electrical circuits is easier and has been more cultivated (for practical purposes) than that of air vibrations, acoustic science profits from electrical by a free exchange of ideas about vibrations. Many acoustical problems can be translated into problems concerning electrical networks, and as there exists a great body of knowledge of such networks, the problem is often solved in the act of translation. Further, by adopting the phraseology of the electrician into acoustics, so that translation of the acoustic problem into the electrical problem becomes automatic, a language for thinking and talking becomes available and is found to clear the mind and assist reasoning. (Eccles 1929, 233)

Furthermore, Eccles observed that, despite the prominence given to the study of acoustics by Rayleigh, the new field of electroacoustics was not studied as eagerly in Britain as it was in the United States and Germany (239).

One of the most significant places for the study of electroacoustics in Germany was the Siemens Research Laboratories. Siemens's most prominent acoustician, Ferdinand Trendelenburg, assembled a monograph on *Die Fortschritte der physikalischen und technischen Akustik* (The proceedings of physical and technical acoustics) in 1932, which he extended in a second edition in 1934 (Trendelenburg 1932, 1934). Trendelenburg's proceedings exhibit the translation of acoustic problems into electrical problems specifically by the use of equivalent circuit diagrams. In the section on hearing and speech, which was Trendelenburg's own research interest, he discussed a schematic representation of speech that used an electric circuit of a self-exciting radio transmitter as an equivalent representation of the human voice (figure 6.2). Trendelenburg referred to two scientists from Bell Laboratories, Irving Crandall and Raymond Lester Wegel, the latter of whom had discussed representing the vibration of the vocal cords by an electric equivalent circuit.[28] John Q. Stewart, a research engineer in the Development and

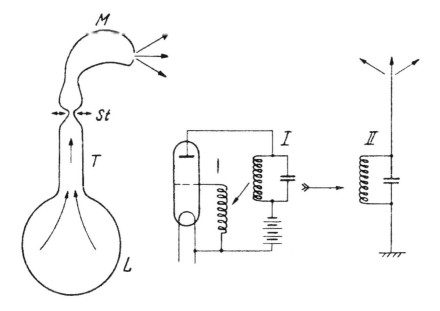

Figure 6.2
The self-exciting radio-tube transmitter as an equivalent circuit for the human voice (Trendelenburg 1934, 76).

Research Department of AT&T, had already published a circuit diagram for producing simple speech sounds in 1922.[29]

Developing equivalent circuits for the human voice was certainly useful if one intended to build an analog electric human-speech synthesizer, an endeavor in which Bell Laboratories became a central actor. Electric-circuit analysis, however, could be applied to a whole range of acoustic systems that would never be built as real electric machines, like the voder, the name under which the human-speech synthesizer came to be known.

Trendelenburg added a separate section on noise abatement in the second enlarged edition of his monograph in 1934. The growing automobile industry in Europe and North America needed to silence the noise produced by combustion engines of cars and motorbikes. Most of the sound silencers were built on the principle of acoustic filters (Trendelenburg 1934, 168). The theory of electric filters had advanced quite far. George W. Stewart had designed an aircraft sound locator for the U.S. National Research Council during World War I (G. W. Stewart 1919a, 19191b). In the early 1920s, he worked on acoustic wave filters based on analogies with well-understood electric wave filters. G. Stewart sold some of his patent rights to Bell Laboratories but did not envision the sound silencer as an application (E. Thompson 2002, 115–168). Martin Kluge, electrical engineer and former student of Heinrich Barkhausen in Dresden, then used electric filter theory to translate the sound silencer into an electric circuit diagram, as I discuss in chapter 4 (see figure 4.1).

Architectural acoustics and the sound insulation of buildings were other prominent areas for the representation of acoustic systems by equivalent circuit diagrams, as we see in chapter 1 (see figure 1.1). Not only Reno Berg and Johan Holtsmark in Trondheim, but also Erwin Meyer, the head of the Department of Acoustics at the Heinrich Hertz Institute in Berlin, used equivalent circuit diagrams in studying how sound traveled through a wall (Meyer 1939, 112). In the 1930s, Meyer gave lectures in the United States, the Soviet Union, and Britain. The five lectures that Meyer gave at the Institution of Electrical Engineers in London in autumn 1937 were subsequently published in 1939 under the title *Electro-acoustics*, one among several monographs on electroacoustics at the time (for example, Le Corbeiller 1934; Gemelli and Pastori 1934).

In the interwar period, the electroanalog understanding of acoustics and its technologies and methods, especially the use of equivalent circuit diagrams, unfolded into a new language of sound. Equivalent circuit diagrams started out as a method to describe and analyze purely electric systems, such as transformers, and were then extended to coupled systems that contained both electric and mechanical components, such as loudspeakers and microphones. Interestingly, many of the acoustic systems represented by equivalent circuit diagrams in the 1930s did not contain any electric components in their material composition, as we have seen in the examples of sound traveling through a wall in a building, the human voice, and automobile sound silencers.

The widespread use and importance of analogies in science can be traced throughout history (Hentschel 2010). The heuristic functions of analogies are various and multilayered. Analogies can propagate a certain worldview by relating elements. They can also serve more practical and pragmatic goals, such as to conceptualize, analyze, and solve scientific and technical problems. Analogies can allow the transfer of thought patterns, language, and mathematical formalism, and can be used to build models and machinery. They can point toward the likeness of phenomena and supply arguments for identifying two apparently different phenomena as being physically identical. This was the case in Maxwell's and Hertz's identification of light as an electromagnetic phenomenon. But instead of claiming physical identity, analogies can also point toward structural identity and aid the visual and functional understanding of systems and processes by transferring from the well understood to the novel and unknown (Hentschel 2010, 32–33).

How, then, did the electrical representation of sound in the 1930s differ from the Maxwellian representation of electromagnetic phenomena through the mechanics of the ether in the late nineteenth century? Victorian physicists like Maxwell, FitzGerald, and Lodge imagined electromagnetic phenomena as mechanical machinery in order to understand and explain their mode of operation. Even though the mechanical models were not thought of as true representations of the ether, the ether was believed to be of a truly mechanical nature. To conceive of a mechanical model meant to have understood the phenomenon. Yet, these mechanical models of electricity did not constitute a structured language that allowed analyzing and solving problems by analogy in the same way as equivalent circuit diagrams did. Despite the efforts of Christopher Polhem and Franz Reuleaux to conceive such a language of machine parts, there was no mechanical alphabet or grammar similar to the equivalent circuit diagram.[30] Some of these models were built as functional devices for lecture demonstration. These demonstration devices not only were meant to show the fundamentally mechanical nature of electromagnetism, through their analogy with well-understood mechanics, they also taught abstract and unfamiliar electromagnetic phenomena to scientists and engineers. Then, with the decline of a mechanical worldview and the increasing familiarity of scientists and engineers with electrical technology, the need to explain the mechanism of electromagnetic phenomena decreased.

The use of electrical analogies and equivalent circuit diagrams in acoustics in the twentieth century was, in contrast, driven not by worldviews but by more pragmatic and technical considerations. With the rapid decline of the electromagnetic worldview, no one claimed physical identity or argued for an foundationally electrical nature of sound. Physical acoustics continued to be understood essentially as a section of the motions of elastic bodies within the field of mechanics as introduced by Helmholtz and Rayleigh. Why did these electrical representations of acoustic systems still become so powerful and widespread in the interwar period?

The answer to this question can be given by three arguments. First, measurement technology in acoustics research had become increasingly electrical. Before the electrification of acoustic measurement, a trained ear and an understanding of the system of European classical music were required. After electrification, the acoustician was not supposed to trust her or his own ear but had to develop an understanding of the design, behavior, and manipulation of electric circuits and transducers. Even if the acoustic system that was analyzed was purely mechanical, like a wall construction or a sound silencer, electrical technology and electrical thinking were already present. Electrical engineers and scientists proficient in electrical measurement technology therefore dominated acoustics research, even in fields like civil engineering and automobile mechanics.

Second, electric oscillations were well understood and structurally analogous to acoustic vibrations. Acoustic systems could therefore be translated into electrical systems. With the concepts of the voltage-source equivalent and the current-source equivalent, circuit diagrams had become a powerful tool to conceptualize, analyze, and solve all kinds of complex oscillation problems, whether they were electrical, electromechanical, or purely mechanical. The language of equivalent circuit diagrams was a language of signs and relations that offered a reduced representation in which extraneous information was eliminated. Dealing with electric systems had become common practice for a large community of scientists and engineers who shared the language of the circuit diagram.

Third, equivalent circuit diagrams were useful for designing, analyzing, and improving technology. This could be electroacoustic technology, for example, a loudspeaker, a microphone, or an audio amplifier. But purely mechanical technology like a wall construction or an automobile sound silencer could also be successfully analyzed and improved with the help of equivalent circuit diagrams. As a structuralist language, it created a direct link between sound and the industrial design and production of technology.

The electroanalog understanding of sound persists today as one of the foundations of acoustics in textbooks and manuals, and as an inherent component of the acoustician's conceptual toolbox. Of course, since the 1960s, a new digital concept of sound has emerged, with the arrival of compact discs, computer music, and MP3 players, which has taken over the analog sound we had gotten so used to from radio, vinyl discs, and cassette tapes. Another translation has taken place from the analog to the digital, where electric signals are converted into bits and stored in binary form. Through digitization, our way to conceptualize, manipulate, and experience sound has again changed significantly. Together with computer algorithms, the concept of information has entered our aural world as a novel but fundamental entity. Again, a new phraseology of acoustics has become automatic "and is found to clear the mind and assist reasoning" (Eccles 1929, 233). What has persisted into the twenty-first century is the essentially cultural as well as technological nature of our approach to sound.

Notes

Chapter 1

1. I have borrowed my title from "Elektroakustikkens tidsalder" (The age of electroacoustics), an article published in 1930 in the Norwegian amateur radio journal *Norsk radio* under the pseudonym Abu Markub (1930).

2. For the history of science's turn toward practice and culture, see among others Pickering (1992) and Buchwald (1995).

3. In her introduction to the *Isis* focus section on science, history, and modern India, Jahnavi Phalkey asks us as historians to critically look at what (Indian) scientists actually do, why they do what they do, and how they argue then act on their arguments, instead of following preconceived ideas of what we think they might be up to (Phalkey 2013, 335).

4. Even today, the university physics curriculum separates the discipline into classical and modern physics, with acoustics seen as a part of classical mechanics. The theory of classical mechanics is assumed to have been completed in the nineteenth century, followed by the development of quantum mechanics and general relativity. When I studied physics at the University of Oldenburg in Germany in the 1990s, the acoustics group was actually one of the largest research groups in the department. Was it all merely applied research derived from a theory that had long been completed?

5. See F. V. Hunt's introduction to the history of electroacoustics (1954) and Beyer (1999).

6. See Gieryn (1983, 1999) for the concept of boundary work in science.

7. For an overview of this literature, see Braun (2004); Sterne (2012); Pinch and Bijsterveld (2012).

8. See, for example, Pantalony (2009); Jackson (2006); Hui (2013); Hiebert (2014).

9. See, for example, Beyer (1999, 177); Lindsay (1945, xxix); Meyer and Waetzmann (1936, 114); and Meyer (1938, 241). Noteworthy among the physicists who did work

on acoustics in the first decade of the twentieth century are Wallace Clement Sabine at Harvard, with his investigations of architectural acoustics; Erich Waetzmann in Breslau (now Wrocław); Max Wien in Danzig (today's Gdańsk); and Chandrasekhara Venkata Raman in Calcutta, with his investigations of musical acoustics. See E. Thompson (2002, 13–113); Beyer (1999, 218–219).

10. See, for example, Beyer (1999, 195–202); Hackmann (1984); Hartcup (1988); Schirrmacher (2009); Barkhausen and Lichte (1920); Waetzmann (1921a, 1921b); Angerer and Ladenburg (1921).

11. Stumpf (1908, 225–227). See Daston and Galison (1992) for a discourse on objectivity in scientific imagery, photography, and what Daston and Galison have called "mechanical objectivity."

12. Barkhausen (1911, 516). See also Sterne (2003) on the cultural encoding of sound inscription of the phonograph and other sound-recording technologies.

13. See Kestenberg (1930). The various contributions were not coordinated but discussed the relationship of art and science from different perspectives.

14. Very little has been written on the history of circuit design. See Jones-Imhotep (2008) for debates in the postwar period, and Ferguson (1992, 11) for abstract visual concepts in engineering. Eugene Ferguson argues that these concepts were specific to engineers, whereas scientists tended to use mathematical concepts. In the history of electroacoustics, however, it is not always possible to separate scientists from engineers, especially in the case of technical and industrial physicists, where many of the actors crossed the line and felt at home in both communities, science and electrical engineering.

15. Jones-Imhotep (2008, 416). It was thanks in large part to standardized electric components that engineers, scientists, and technicians could build material circuits from a circuit diagram. Radio amateurs built their own equipment from circuit diagrams published in journals such as *Wireless World* and components from radio supply shops.

16. Berg and Holtsmark (1935b). Johan Holtsmark was professor of physics at the Norwegian Institute of Technology in Trondheim, where he established an acoustic laboratory in 1929. His assistant, Reno Berg, was an electrical engineer. See Wittje (2003, 69–142).

17. See, for example, Trumpy (1930, 13–14, 52–53).

Chapter 2

1. Science education was also expanded in German secondary schools, the *Gymnasien*, in the German Empire after 1871. The traditional humanist gymnasium, which favored classical languages and humanistic subjects over science, was rivaled

by the new *Realgymnasium* and the *Oberrealschule*, which prioritized modern languages and science. The heterogeneous system of technical education expanded, further increasing the need for science teachers in vocational schools and technical colleges (see König 1993).

2. This tension between scientific method and practical entrepreneurship in electrical engineering and industry has been discussed in detail for the cases of Thomas Edison, Guglielmo Marconi, and the "British electrical debate" between William Preece and Oliver Heaviside (see, for example, Hong 2001, 36–41).

3. In 1884, Helmholtz became related to Siemens by marriage when Siemens's eldest son, Arnold, married Helmholtz's daughter Ellen (Cahan 1989, 37).

4. To avoid confusion, I will refer to William Strutt, 3rd Baron Rayleigh, simply as Rayleigh.

5. Helmholtz argued that the aesthetic perception of music was much more directly related to the sensations of tone than the visual arts, which had a rather indirect relationship with the sensations of light (Helmholtz 1863, 4).

6. My translation. "So haben also die Wissenschaften einen gemeinsamen Zweck, den Geist herrschend zu machen über die Welt. Während die Geisteswissenschaften direct daran arbeiten, ... das Reine vom Unreinen zu sondern, so streben die Naturwissenschaften indirect nach demselben Ziele, indem sie den Menschen von den auf ihn eindrängenden Nothwendigkeiten der Aussenwelt mehr und mehr zu befreien suchen." See also Heidelberger (1994, 174).

7. Helmholtz was personally connected to Fichte through his father, a close friend of Fichte's son Immanuel Hartmann von Fichte.

8. From Helmholtz's "Plan of Work" in his 1863 *Sensations of Tone*.

9. Translation modified.

10. Wave formalism allowed the transfer of formalism from acoustic wave theory to quantum mechanics; thus Hilding Faxén and Johan Holtsmark, for example, transferred Rayleigh's wave scattering to electron scattering in 1927 (Faxén and Holtsmark 1927; Wittje 2003, 145–146).

11. Beetz argued that his tuning fork apparatus was both cheaper than a clockwork arrangement and more reliable in operation because of its simplicity. He introduced the instrument to measure the velocity of artillery shells but also proposed physiological measurements as an application (Beetz 1868). In 1911, the instrument company Max Kohl gave the accuracy of the tuning fork chronograph (as suggested by Beetz) as within 0.0005 seconds (Max Kohl AG 1911, 2:248).

12. The world of European classical music itself was unstable throughout the nineteenth century (Hui 2013, 23). Despite the continued conviction that German music

was superior to all others, musicians and musicologists became increasingly aware of the historicity of musical taste and music perception.

13. On Koenig, see Pantalony (2009).

14. The translation of Koenig's notion of Stoßton as "beat note" is confusing since it is too close to the notion of "beats," the English translation of Helmholtz's concept of *Schwebung*.

15. Stumpf directly compared the recording of exotic music to the visual recording of the faces of exotic people. Before photography, Stumpf argued, illustrators Europeanized the features of "savage people" because they were not able to see them objectively. It was not the eye but the brain that guided the illustrator's crayon. Only photography allowed a truly objective recording of these faces (Stumpf 1908, 226).

16. See also E. Thompson (2002, 132–144); Bijsterveld (2008, 137–158).

17. Luigi Russolo to Francesco Balilla Pratell, 11 March 1913, first published in 1916 in *L'arte dei Rumori*.

18. Jonathan Sterne interprets both the telephone and the phonautograph as tympanic mechanisms, since their membrane mechanism was modeled on the tympanic membrane in the human ear; see Sterne (2003, 22–23, 77–81) for the connection between Helmholtz's physiology of hearing and the inventions of the phonograph and telephone.

19. David Edward Hughes was an inventor and musician. He was professor of philosophy and music at a college in Kentucky. More an inventor and engineer than a scientist, Hughes developed a telegraph instrument with a keyboard, which was in widespread use by the 1870s. In 1877, Hughes moved to London, where he experimented with the Reis telephone and his microphone. Hughes has been credited with having produced electromagnetic waves before Hertz, as early as 1879. His claim to the novelty of the effect he produced was refuted by Gabriel Stokes when Hughes demonstrated his experiments at the Royal Society in 1880. In contrast to Hertz, Hughes did not have the mathematical training to understand Maxwellian electrodynamics and connect it to his experiments. See I. Hughes (2009, esp. 125, fig. 8) for Hughes's induction balance and sonometer.

20. Heaviside's work led to the "British electrical debate," a conflict between the Maxwellians, who proposed loading the telephone lines with additional self-induction, and William Henry Preece of the Post Office Telegraph Department as well as other practitioners, who opposed such loading (B. Hunt 1991, 129–151).

21. For a history of the origins of the equivalent circuit concept, see also D. H. Johnson (2003a, 2003b).

22. The full title is *Über den Einfluß der Amplitude auf Tonhöhe und Decrement von Stimmgabeln und zungenförmigen Stahlfedern. Elektroakustische Untersuchungen* (Hartmann-Kempf 1903).

23. Based on his dissertation, Robert Hartmann-Kempf filed several patents on a frequency meter and a speed indicator for Hartmann and Braun, which produced the frequency meter as System Hartmann-Kempf.

24. See Hartmann and Braun (1886) and the Hartmann and Braun catalog (1894, 37, 58–60).

25. In his publications of the 1890s, Wien did not use the term "elektroakustisch," which was only introduced after 1900. I therefore prefer to refer to his practices as "electrical and acoustic."

26. The electromagnetic worldview reached its zenith about 1905 and lost its appeal around 1914 with the rise of quantum theory and relativity. See Kragh (1999, 103–119).

27. On Poulsen, see Hong (2001, 165–169). Poulsen also developed and produced a magnetic wire recorder for audio recording, the Telegraphone, for which he obtained a patent in 1898 and which he presented at the World Fair in Paris in 1900.

28. Among Poulsen's competitors were the Marconi system, Telefunken's Lösch-funkensender (developed by Max Wien around 1905), and machine alternators, which also produced undamped continuous waves.

29. See des Coudres (1919). Like Max Wien, Simon had studied with August Kundt at the University of Berlin. He was then the assistant of Eilhard Wiedemann in Erlangen and Eduard Riecke in Göttingen.

30. See "Bericht über die Thätigkeit der Abteilung für angewandte Elektricitätslehre im Jahre 1902/03," p. 3, Protokoll d. Generalvers. d. Göttinger Vereinigung vom 28 November 1903, SUB Göttingen Math. Archiv, 50:11.

31. See Manegold (1970, 85–244) on Felix Klein and the foundation of the Göttinger Vereinigung. In 1900, the name of the association was extended to Göttinger Vereinigung zur Förderung der angewandten Physik und Mathematik (Göttingen Association for the Advancement of Applied Physics and Mathematics). The Göttinger Vereinigung organized meetings, exchanges, and excursions to factories.

32. Historian of the life sciences Hans-Jörg Rheinberger introduced the concept of the experimental system as a "basic unit of experimental activity combining local, technical, instrumental, institutional, social, and epistemic aspects" (Rheinberger 1997, 238). The electric arc at Simon's Institute for Applied Electricity at the University of Göttingen meets all these criteria.

33. Gertrud Lange was the only woman in the group of Simon's students on the electric arc. She had studied for several semesters in Glasgow and in Göttingen before she was allowed to graduate with the *Abitur*, the German university entrance qualification, in Hannover in 1906. For her doctoral thesis in 1909, she studied the hysteresis of the electric arc (G. Lange 1910; Tobies 1997, 34).

34. Lichte came to lead the development of the German system for sound motion pictures in the 1920s. Rihl worked at the Siemens laboratories.

35. See "Bericht des Herrn Professor H. Th. Simon über die Tätigkeit des Instituts für angewandte Elektrizität 1906/7," p. 72, Protokoll d. Generalvers. d. Göttinger Vereinigung vom 5–6 Juli 1907, SUB Göttingen Math. Archiv, 50:19.

36. See "Bericht des Herrn Professor Dr. Simon über die Tätigkeit des Instituts für angewandte Elektrizität im Jahre 1907/08," p. 72, Protokoll d. Generalvers. d. Göttinger Vereinigung vom 16 u. 17 Okt. 1908, SUB Göttingen Math. Archiv, 50:21.

37. Stefan Wolff has suggested that there was little or no organized contact between university physics departments and the military, and that it was left to the initiative of the physicists to put their department at the service of the war effort (Wolff 2006). This was certainly not the case for the University of Göttingen's Institute for Applied Electricity. In 1908, the year that Simon set up the Radioelektrische Versuchsanstalt, Ludwig Prandtl set up the Modellversuchsanstalt, a model testing facility with a wind tunnel. While the Radioelektrisch Versuchsanstalt was funded directly by the German Imperial Navy, the Modellversuchsanstalt was funded by the Motorluftschiff Studiengesellschaft (Association for the Study of Motorized Airships). The association, which was headed by former secretary of the German Imperial Naval Office Friedrich von Hollmann, brought together industrialists and representatives of the military. See Busse (2008, 55–58); Eckert (2006, 45–48).

Chapter 3

1. Trischler (1996, 96). The crisis of the German physics community in the aftermath of World War I is discussed in the context of the Forman thesis in Meyenn (1994) and, more recently, by Carson, Kojevnikov, and Trischler (2011).

2. Beyer (1999, 197), Lindsay (1945, xxix). See also Meyer and Waetzmann (1936, 114). Erwin Meyer, born 1899, was conscripted as a soldier in 1917 and enrolled as a student only in 1918 (Guicking 2012). Meyer's mentor for his doctoral dissertation, the Breslau professor of physics Erich Waetzmann, however, could draw on his own experience. Waetzmann had worked on sound ranging of aircraft and the development and use of a geophone for the acoustic location of tunnels dug under trenches (Waetzmann 1927; Meyer 1938; Waetzmann, "Praktische Erfahrungen über das Abhören von Fliegergeräuschen," 3 May 1918, Bundesarchiv-Militärarchiv [BArch] PL 2-IV/323).

3. Ernest Lancaster Jones collected the objects for the Sound Department, along with substantial documentation material (Lancaster Jones 1922; Science Museum Archive [SciMus], MS 2019). Correspondence and registration numbers suggest that all the objects were collected in 1921, apart from the Tucker microphone, which was added in 1922. While the artillery-ranging apparatus is quoted as being presented by the Imperial War Museum, the Admiralty donated the hydrophones (SciMus MS 2019/53). The hydrophones and the artillery-ranging apparatus are still part of the Science Museum collections, though I was not able to find the aircraft locator mentioned in Lancaster Jones's 1922 article. American hydrophones from World War I were exhibited at the Smithsonian Institution around 1920 (Hackmann 1984, plate 3.8).

4. Whereas technical physics was struggling to become institutionalized as a degree subject at the Technische Hochschulen around the turn of the century, chemistry programs were part of the Technische Hochschulen from their very beginning as an institutional category. Prior to World War I, only the Technische Hochschule Munich had a laboratory for Technische Physik, which offered degrees from 1902 on (Knoblauch 1941–1942).

5. In 1920, the foundation was renamed the Kaiser-Wilhelm-Stiftung für technische Wissenschaft (Kaiser Wilhelm Foundation for Technical Science); it was finally dissolved in 1925. Manfred Rasch attributes the failure of the foundation partly to "the difficulty of achieving academic status for the technical sciences in Germany" (Rasch 2006, 192). However, it contrasts with the overall success of technical physics in Weimar Germany, in the course of which the rhetoric of technical physics drew heavily on the continuation of practices from World War I.

6. The University of Jena established an extraordinary professorship in technical physics with the support of the Zeiss Foundation in 1897. Max Wien was one of the country's main representatives of technical physics. Before his appointment in Jena, he had been professor of physics at the Technische Hochschule Danzig, where he had developed the Löschfunkensender, the radio transmitter system used by Telefunken (Gerber 2009, 159–160).

7. See Kaufmann (1996, chapter 3, 170–261) for a media history of telephony, telegraphy, and wireless during World War I.

8. My translation. "In den Kriegsjahre [des Ersten Weltkriegs], als draußen die Geschütze donnerten, war es umso stiller auf den Straßen und in den wissenschaftlichen Instituten Berlins. Fast wie in einem schalldichten Raume ließen sich akustische Beobachtungen durchführen und selbst die Flüsterlaute in ihre letzten Bestandteile zerlegen."

9. I describe these military systems as large technological systems in Thomas Hughes's definition (T. Hughes 1987, 51–56).

10. As far as I can tell from the sources, sound observers on the battlefields of World War I were male. Behind the front line, however, psychologist Emily Mary Smith developed aural tests during the war to select hydrophone operators at the Cambridge Psychological Laboratory (Smith and Bartlett 1919, 1920).

11. See Harbeck (1943a, 6); Schirrmacher (2009, 169). Among Born's contributions to sound ranging were meteorological corrections, accounting for the effect of the wind gradient on the trajectory of the sound (Angerer and Ladenburg 1921, 314–317).

12. See H. Schubert (1987) for Mandelung; Jung (1961) for Fredenhagen; Joos (1953) for Angerer. Interestingly, all three had been in Göttingen around the same time and had most likely known each other ever since. Mandelung was Simon's doctoral student and Fredenhagen Simon's assistant, but I know of no such connection between Simon and Angerer.

13. Lieutenant R. Harbeck evaluated time measurement with microphones and an oscillograph as being ten times more precise than measurement with human observers and stopwatches (Harbeck 1943b, 10). Harbeck's detailed report on German Army sound and flash ranging during World War I must be read with great caution. Being explicitly anti-Semitic, Harbeck blamed the apparent failure of the German Army to develop an objective sound-ranging system on "the number of Jews employed," specifically naming Ladenburg and Born (Harbeck 1943b, 9). In the main report, Harbeck did not mention the names of any Jewish contributors (which included Ladenburg, Born, Löwenstein, Hornbostel, and Wertheimer), but he discussed their work in detail as "German contributions" (Harbeck 1943a). Richard Berger gave the precision of the subjective method with stopwatches as within a range of 40 to 60 meters (Berger 1918, 9). Harold Winterbotham noted that "long training and experience are necessary before any reliable results can be obtained" using the stopwatch sections (Winterbotham 1918, 33). In contrast, a British report found that under favorable conditions, an accuracy of 25 yards (23 m) could be guaranteed for the British objective system (*Sound Ranging* 1917, 6).

14. See C. Hoffmann (1994, 266). Along with Kurt Koffka, Köhler and Wertheimer were cofounders of the school of Gestalt psychology, while Hornbostel belonged to the larger circle of Gestalt psychologists. For Köhler, see Bergius (1979).

15. It is difficult to estimate how many sections used the binaural system as opposed to stopwatches. According to Berger, the binaural method was not very precise and was useful only in the hands of specially trained observers who were able to determine the direction to "a few sixteenths of a degree" (Berger 1918, 17–18). The Nazi propaganda writer Martin Bochow, in his autobiographical *Schallmesstrupp 51*, reported that he was introduced to the binaural method in his sound-ranging section in 1916 and only later to the method measuring time differences with stopwatches (Bochow 1933).

16. Hornbostel and Wertheimer sound locator patent record: Patentschrift Reichspatentamt 301669 Klasse 74d Gruppe 5, filed 7 July 1915, issued 28 September 1920, DEPATISnet, Deutsches Patent- und Markenamt patent database, https://depatisnet.dpma.de/DepatisNet/.

17. The official British *Report on Survey on the Western Front 1914–1918* (Jack 1920, 111) identified the microphones as the Paris-Rouen model.

18. Angerer and Ladenburg (1921, 295, 319); Angerer (1922); Harbeck (1943a, 15). For Zenneck's observations on the battlefield, see Zenneck (1960, 297–300).

19. Harbeck mentions a Siemens and Halske microphone for the hearing impaired and an apparatus that Siemens and Halske had set up for investigations at the Berlin subway in 1913 (Harbeck 1943a, 37).

20. The manuals were translated into German and their content investigated. See "Bericht zu den erbeuteten Merkblättern über das englische Schallwesen," 15 pp., BArch PH 3/506. The original British manual was dated March 1917.

21. Allgemeines Kriegsdepartement, 18 December 1917 and 19 December 1917, BArch PH 3/506. The Ministry of War had already instructed army scientists to conduct experiments on these registration methods, in June 1917, and had ordered five oscillographs, five string galvanometers, and ten smoked paper recorders for that purpose (Kriegsministerium, 26 June 1917, BArch PH 3/506).

22. Artillerie-Messschule Wahn, 3 January 1918 and 6 January 1918, BArch PH 3/506.

23. Allgemeines Kriegsdepartement, "Vorläufiger Bericht über das erbeutete englische 'Mikrophon,'" 15 April 1918, BArch PH 3/506.

24. Waetzmann, "Praktische Erfahrungen."

25. Waetzmann (1912); see also Gibling (1917) for a comparable account on musical listening.

26. Waetzmann, "Praktische Erfahrungen," p. 1.

27. Ibid., 2–3.

28. Ibid., 4–5.

29. Ibid., 8.

30. Ibid., 10.

31. Ibid., 12.

32. Ibid., 14.

33. Van der Kloot (2011) and the archive collection SciMus MS 2019.

34. C. Jakeman, "Note on the influence of the separation of the receivers upon the accuracy of binaural sound locators," 5 November 1918, p. 1, SciMus MS 2019/18; and Bloor (2000, 203).

35. Comparison by Jay Hopkins, Chief of the American Anti-Aircraft Service, undated, MS 2019/34; A. V. Hill, "The Perrin telesitemeter, the Baillaud paraboloid, and the ring sight," 7 October 1918, MS 2019/30; and A. Ward's visit to Pont-sur-Seine, 10 August 1918, MS 2019/23, all in SciMus.

36. "The Use of Long (Four to Six Meter) Horns in Airplane Direction and Location," report undated but attached frequency curve dated 26 July 1918, SciMus MS 2019/24-335; see also G. W. Stewart (1919a, 1919b). See Bloor (2000, 204) for an image of a U.S. sound detector with four large exponential horns.

37. "Summary of a Report on a Pair of Darwin Listening Trumpets by Admiral Sir Henry Jackson R.N. F.R.S.," 1 March 1918, SciMus MS 2019/37.

38. "Note on Sound Location by Double Disc," 8 June 1918, SciMus MS 2019/14a.

39. Tucker, "Report on Sound Mirrors," 10 October 1918, SciMus MS 2019/6. See also Zimmerman (2001, 7–10), especially for the Mather mirrors. Tucker experimented with the French Baillaud paraboloid as well.

40. Tucker, "Report for week ending March 16th, 1918," SciMus MS 2019/3.

41. Tucker, "Report for the MID 18 Feb. 1918," SciMus MS 2019/2. The Mark II WT amplifier probably refers to the Mark II wireless telephone used by the Royal Flying Corps since 1917.

42. Tucker, "Report for the Week Ending April 6th/18," SciMus MS 2019/5.

43. Tucker to the Controller, Munitions Inventions Department, 10 February 1918, p. 10, SciMus MS 2019/2.

44. The German Torpedo-Versuchs-Kommando experimented with wireless underwater communication, referred to as *Stromlinientelegraphie*, as well as with methods of detecting submarines electromagnetically under water. Arnold Sommerfeld carried out calculations on the feasibility of different proposals for electromagnetic underwater telegraphy, for which he met with Heinrich Barkhausen in May 1917 (Sommerfeld, "Für den Jahresbericht der K. W. K. W.," March 1918, in Eckert and Märker [2000, 589–591]). All experiments and calculations showed that because of the high conductivity of seawater, all electromagnetic effects were absorbed at relatively short range. The most sensitive laboratory equipment gave a maximum range of 150 meters for electromagnetic detection, which, of course, also meant that the German U-boats did not need to fear such detection from the enemy. Tätigkeitsbericht der Abteilung V.K.U., undated, BArch RM 27-III/29.

45. Walter Hahnemann for Signal Gesellschaft to the Staatssekretär des Reichs Marine Amtes, 20 February 1918, BArch RM 5/33.

46. Eduard von Capelle, "Betr: Denkschrift der Signalgesellschaft," 14 March 1918, BArch RM 5/33. See BArch RM 120/95, 15 July 1915, on the navy evaluation of the underwater telegraphy systems delivered by the Signal Gesellschaft and the need to train personnel.

47. Gerdien and Riegger (1920); Boedeker and Riegger (1920); Gerdien (1922). See Aigner (1922, 206) for the Siemens signal receiver, (227) for the Siemens noise receiver, and (234) for a Siemens resonance telephone. Among the patents that Riegger and Gerdien took out for Siemens are Einrichtung zur Schallübertragung unter Wasser, DRP 357651, Klasse 74d, Gruppe 6, 14 April 1916, and Einrichtung zur elektrostatischen Erregung einer Membran, vorzugsweise zur Erzeugung von Schallwellen, DRP 387116, Klasse 74d, Gruppe 5, 18 May 1918, DEPATISnet.

48. Barkhausen had used the transmitters as receivers alongside the carbon microphones because the microphones exhibited large variations in sensitivity, making quantitative measurements difficult (Barkhausen and Lichte 1920, 486, 497).

49. My translation. "Während der Tonempfänger zur normalen und störungsfreien Aufnahme von Unterwasserschallsignalen dient, schuf der U-Bootkrieg das Bedürfnis nach einem zweiten Empfängertyp, der hauptsächlich dazu bestimmt war, Geräusche von Schiffsschrauben oder fremden U-Booten aufzunehmen und daher den Namen Geräuschempfänger erhielt. … seine Eigenschwingungen [sind] sehr stark gedämpft. Dadurch wird er eben befähigt, Bordgeräusche anderer in der Nähe befindlicher Schiffe gleichmäßig aufzunehmen, wie das Arbeiten der Schiffsmaschinen, Pumpen, elektrische Maschinen und das Mahlen der Schiffspropeller. Alle diese Geräusche haben für den Hörenden eine charakteristische Färbung, so daß Uebung in die Lage versetzt, aus den Geräuschen die Schiffsgattung, die Zahl der Schrauben und ihre Flügelzahl, die Art des Schiffsantriebes, ob Kolbenmaschine oder Turbine usw. mit voller Sicherheit anzugeben." Aigner was scientific assistant to Gustav Jäger in the physical laboratory at the Technische Hochschule in Vienna. During World War I, he was an adviser in underwater acoustics for the Austro-Hungarian Navy, as an extension of the research carried out by the German Navy (Kurzel-Runtscheiner 1953). According to Aigner, the Austro-Hungarian efforts did not get very far before the end of the war (Aigner 1922, 68). A German Navy report of 13 January 1915 on the different sound receivers, entitled "Unterwasserschall-Empfangsanlage für die UB- und UC Boote," can be found in BArch RM 120/95, 44–45.

50. "Niederschrift über die Besprechung am 17. Mai 1916 betr. U-Bootston" and "Im Anschluß an U.J. G.12822 betr. U-Bootston," 24 May 1916, BArch RM 5/33 3516.

51. Inspektion des Torpedowesens B. Nr. Gg 650 VK U., 3 June 1916, BArch RM 5/33 3516. See Hackmann (1984, 60) for the British towing hydrophones, and (63) for the British assessment of captured German equipment.

52. See Rössler (2006, 23–24) for a detailed account of the orders to silence submarines.

53. Schmaltz (2013, 18). See also Eckert (2006, 163); Rotta (1990, 168).

54. See "Kommando der Hochseestreitkräfte G. 4203 F.2.," 20 May 1918, for the Heligoland station, and "Betrifft: Feste U G E-Anlagen," 2 July 1918, for the Zeebrugge and Ostend stations, in BArch RM 5/23 3518. For the technical details of the stations, see "Bestimmung des Ortes einer Schallquelle," undated; "Betrifft: Warnanlage Zeebrügge, Ostende," 20 July 1918; "Niederschrift der Besprechung vom 15. VII 1918," 15 July 1918; and "Anträge für Einbau des Suchgerätes und der Warnanlage an der flandrischen Küste," undated, all in BArch RM 120/60. It is not clear whether the oscillograph was used before the war ended.

55. For the development of asdic and the history of piezoelectricity, see Hackmann (1984, chapter 6) and Katzir (2006, 2010, 2012).

56. Adler and Butler (1999, 105–107); Hackmann (1984, 37, 61). The Science Museum in London has a selection of British hydrophones collected by Lancaster Jones in 1921, among others the two fish hydrophones with the inventory numbers 1921-60 and 1921-61.

57. Kevles (1978, 119–121). As in the German and British case, there were tensions between different actors during the American development of listening devices for underwater acoustics. In the United States, these tensions arose between Edison, along with the Naval Consulting Board, which favored a practical engineering approach, and the scientists of the National Research Council, headed by George E. Hale and Robert A. Millikan, which favored a scientific approach (Kevles 1978, 102–138).

58. Willem Hackmann places both incidents in the year 1916 (Hackmann 1984, 63). However, the examination of *UC 44* could not have happened before 1917. *UC 44* sank on 4 August 1917 on one of her own mines on the coast off Waterford, Ireland, and was raised by the Royal Navy in September 1917.

59. Waetzmann (1921a, 166). See also Waetzmann's report to the German Army, "Praktische Erfahrungen über das Abhorchen von Fliegergeräuschen," abridged copy, 3 May 1918, 14 pp., BArch RL 2-IV/323. Waetzmann's prewar work in acoustics was explicitly in the tradition of Helmholtz (Waetzmann 1912).

60. Waetzmann stated that he had never had the difficulty in analyzing sounds that he had experienced with aircraft sound. Not even the ringing of bells, which he considered challenging, came close (1921a, 166).

61. Physiologist Philipp Brömser of the University of Munich participated in the war as an army doctor. Interested in physical measurement, he designed listening devices for acoustic surveillance (Reichel 1955).

62. Barkhausen, Schaltung zur Verminderung von Störungen beim Empfang von Telephonströmen, DRP 339925, filed on 7 June 1917, issued on 20 August 1921. Barkhausen had patented another circuit to reduce noise in telephone systems without amplification in 1916 (DRP 339924, filed on 8 December 1916, issued on 20 August 1921, DEPATISnet).

63. See Sommerfeld, "Für den Jahresbericht der K. W. K. W.," March 1918, in Eckert and Märker (2000, 589–591), and Tätigkeitsbericht der Abteilung V.K.U., undated, BArch RM 27-III/29.

64. In Germany, Waetzmann, Hornbostel, and Wertheimer came from acoustics research. While Max Wien and Jonathan Zenneck had worked in both fields, Zenneck was seen as a specialist in wireless rather than in acoustics. Other scientists and engineers (including Riegger, Hahnemann, and Barkhausen) came to wartime research and development with clear credentials from wireless only. In Britain and in the United States as well, many scientists and engineers (including Rutherford, Ryan, and Fessenden) had backgrounds in wireless rather than acoustics research and development.

65. In his *Principles of Electric Wave Telegraphy*, John Ambrose Fleming called this the "telephonic method of receiving" (Fleming 1906, 365).

66. See especially Trendelenburg (1927, 2) and Schuster, "Grundbegriffe der technischen Akustik: Das Schallfeld" (1934, 33–46).

67. Hahnemann and Hecht (1919a, 1919b, 1919c). Hecht later published two monographs, on equivalent circuits and differential equations of mechanical and electrical oscillating systems in 1939, and on the theory of electroacoustic transducers in 1941 (Hecht 1939, 1941).

68. Steinmetz and Berg (1897, 184–185); Hahnemann and Hecht (1919a, 1919b, 1919c). Hahnemann and Hecht made no reference to Steinmetz, but they did refer to Gisbert Kapp's 1907 textbook on transformers, which also used equivalent circuit diagrams (Kapp 1907).

69. See C. Hoffmann (1994) as an example of this approach. Christoph Hoffmann's article focuses on the activity of several of Carl Stumpf's disciples during World War I and its significance for the transformation of experimental psychology in Germany after the war.

70. See Hackmann (1984, 64–70) for a detailed record and evaluation of the use of hydrophones against U-boats. The number of German U-boats detected by hydrophones and destroyed was small compared to the expense of the program. The German U-boat campaign ultimately failed in its objectives because of the convoy system introduced by the Allied navies in 1917. The Allied use of hydrophones nevertheless created an effective deterrent, forcing the German Navy to take elaborate measures to silence their U-boats. David Zimmerman finds that the development of

mirror technology for long-range aircraft detection up to 1935 was a failure. Radar replaced long-range acoustic detection. But the final blow for long-range acoustic detection came not from radar alone but from increasingly fast aircraft: when aircraft moved closer to the speed of sound, too little warning time was left using acoustic detection (Zimmerman 1997, 2001).

71. Emily Thompson has discussed Wallace Sabine's use of a telephonic receiver and an "Einthoven string dynamometer" (which must have been an Einthoven string galvanometer) in measurements on the distribution of sound intensity in 1910. In Sabine's 1912 publication, however, he did not mention his measurement setup. Thompson suggests that he was not comfortable with the new technique (E. Thompson 2002, 167–169). Sabine's colleague at Harvard University, George W. Pierce, used Sabine's telephone receiver and a galvanometer in sound measurements in 1910. Pierce found the sensitivity of this detector very weak. He had to tune his telephone circuit to a single frequency and could make only qualitative measurements (E. Thompson 2002, 83–84).

72. Richard Berger was probably the first acoustician to work on noise abatement in Germany. He completed his dissertation on sound transmission through walls with Oscar Knoblauch at the Technische Hochschule Munich in 1911. Berger worked on sound ranging during World War I (Berger 1911, 1926, esp. 56–57).

73. Frank Gray, "Sound-proof headpiece," 29 October 1918, SciMus MS 2019/31 377, and "Aircraft Locating Trumpets," Anti-Aircraft Experimental Section, undated, presumably 1918, SciMus MS 2019/36.

74. Smith and Bartlett (1919, 1920). Emily Mary Smith and Frederic Charles Bartlett married in 1920.

75. Smith and Bartlett (1919, 105, figs. 1 and 2; 106, fig. 3).

76. Fritz Haber was arguably the most prominent advocate of a continuation of military-scientific research practices of World War I into the interwar period. See Haber, "Die Chemie im Kriege," 11 November 1920, in Haber (1924, 25–41). Stumpf argued for the continuation of the military-scientific collaboration in psychology (C. Hoffmann 1994, 274–276).

Chapter 4

1. See Trendelenburg (1927); Waetzmann (1934a). *Handbuch der Physik* appeared in twenty-four volumes between 1926 and 1929, *Handbuch der Experimentalphysik* in forty-five volumes and subvolumes between 1926 and 1937. As the names suggest, the *Handbuch der Physik* covered both experimental and theoretical physics, including atomic physics, quantum mechanics, and the theory of relativity, while the *Handbuch der Experimentalphysik* was limited to experimental physics. Christina

Lembrecht has argued that the *Handbuch der Physik* was dominated by physicists located in Berlin who favored the new theoretical physics, while physicists from south and southeast Germany, who were critical of, if not opposed to, the new theoretical physics, dominated the *Handbuch der Experimentalphysik*. Philipp Lenard and Johannes Stark, the main representatives of the anti-Semitic "Deutsche Physik," were among the authors of the *Handbuch der Experimentalphysik*, so the work acquired a reputation as the *arisches Handbuch* (Aryan handbook). See Lembrecht (2007, 179–184, esp. 184). While I support Lembrecht's general characterization of the two handbooks, it should be noted that the two volumes of *Technische Akustik* did not exhibit a bias against scientists from Berlin, who constituted the majority of the authors, or exclude Jewish contributors. Although the volume was published in 1934, a year after the Nazi seizure of power, one of its authors was Jewish. Hans Salinger, head of the department of telegraphy and measurement at the Heinrich Hertz Institute in Berlin, wrote the chapter on long-distance transmission of sound on cables (Salinger 1934). In 1936, Salinger was dismissed from his position after the reorganization of the Heinrich Hertz Institute, and he then emigrated to the United States (Gundlach 1978).

2. Meyer completed his thesis at the University of Breslau on the effect of tone waves on resonating membranes (Meyer 1923). Trendelenburg wrote his dissertation on the thermophone, an electroacoustic transducer that transformed heat from a conducting wire into acoustic vibrations, under the supervision of Max Reich at the Institute for Applied Electricity (Trendelenburg 1923).

3. For the 1934 handbook, Meyer wrote on measurement technology and on gramophone and magnetic recording, while Trendelenburg wrote on the analysis of internal sounds of the human body. Hermann Backhaus, Hugo Lichte, and Ernst Lübcke also wrote for both handbooks.

4. The National Research Council Committee on Acoustics (1922). I would like to thank Emily Thompson for pointing out this document to me.

5. I have made this case in detail for Johan Holtsmark's establishment of an acoustics laboratory at the Norges Tekniske Høgskole in Trondheim, Norway. Because Norway was neutral during World War I, scientists had not been drawn into the war effort, but the arrival of radio broadcasting and sound motion pictures created a demand for acoustic research and acoustic expertise there as well (Wittje 2003, 69–156).

6. See Serchinger (2008) for an extensive biography of Schottky, especially 588–590 for Schottky's motivations to work at Siemens rather than as a professor in a university department. Schottky is mostly known for his work on atomic theory, amplifier tubes, and solid-state physics, but he also worked on electroacoustics. His work on amplifiers and amplifier tubes and noise in electric systems created a link between these fields.

7. See Forman (1971, 1973); Meyenn (1994); and, most recently, Carson, Kojevnikov, and Trischler (2011).

8. Hoffmann and Swinne (1994, 24–25, 30). See also Richter (1972, esp. 10–12); Gehlhoff (1921, 121).

9. Hoffmann and Swinne (1994, esp. 16–17); Gehlhoff, Rukop, and Hort (1920). Gehlhoff had been Jonathan Zenneck's assistant in Danzig, where he had also completed his habilitation. On Gehlhoff, see Fritz (1964).

10. Gehlhoff, Rukop, and Hort (1920, 4). Helga Schultrich comes to similar conclusions. She estimates the total number of physicists in Germany in 1930, excluding school teachers, as about 1,600, of whom 500 worked in universities and other institutions of higher learning, just over 100 in public research institutions, 100 in the patent system, and about 800 in industry. Schultrich (1982, 217).

11. See, for example, Gehlhoff, Rukop and Hort (1920); Gehlhoff (1921); Krüger (1921). The protagonists of Technische Physik thus carried out extensive "boundary work," to use the term coined by Thomas Gieryn, to set their own disciplinary map against that of general physics as practiced at German universities at the time as well as that of engineering (Gieryn 1983, 1999). For a history of the German Technische Hochschulen and the engineers' boundary work in the demarcation from science, see Manegold (1970).

12. It would distract from the purpose of this book to enter into the general debate on the relationship between scientific research, technological development, and industrial production. See Edgerton (2004) for a critical discussion of the "linear model" in the historiography of science and technology.

13. Arnold Sommerfeld, for example, was one of the protagonists of atomic theory as well as a frequent contributor to *Zeitschrift für technische Physik*.

14. Reiher's and Barkhausen's credibility within the physics community is discussed below (see 4.2.1. and 4.4.2) and in chapter 5.

15. Membership numbers are given in Hoffmann and Swinne (1994, 16). In 1924, the Society for Technical Physics had about 1,660 members, compared to about 1,300 in the Physical Society.

16. See Hoffmann and Swinne (1994, 18–20). For the debate on Technische Physik after World War II, see Ramsauer (1949, 34–47); Vieweg (1948). The chairs, curricula, and departments of technical physics did, however, survive in many German-speaking and other European technical universities.

17. See, for example, Zenneck (1960, 396–399).

18. Reich was not the search committee's first choice as the successor to Simon and was appointed only after several other candidates proved unavailable, among them Max Wien and Heinrich Barkhausen ("Ersatzvorschläge für Professoren,"

17 April 1904–17 December 1920, 282–283 and 326–327, Secreta Kur 4091, Universitätsarchiv Göttingen). That the Institute for Applied Electricity did not regain a central position in electroacoustics research after the war probably had less to do with the person of Max Reich than with the fact that so many other actors had entered the field. Twenty-seven doctoral theses were completed at the institute between 1905 and 1919 under Simon and twenty-eight between 1920 and 1928 under Reich, suggesting that Reich was just as productive as Simon as a teacher (Institut für Angewandte Elektrizität, "Personal" and "Wissenschaftliche Arbeiten," 21 May 1929, Bewilligungen 1905–1934, Kur 1458, Universitätsarchiv Göttingen). Ferdinand Trendelenburg, who graduated in 1922, was among Reich's students.

19. In 1912, Waetzmann became professor extraordinarius at the University of Breslau, where he had also studied, and full professor at the Technische Hochschule Breslau in 1926 (see Meyer 1938). Lübcke set up the acoustics laboratories at the Physics Department of the Technische Hochschule Braunschweig after he became Privatdozent (adjunct professor) there in 1929 (Lübcke 1931). Backhaus was appointed professor of theoretical electrical engineering and low-voltage engineering at the Technische Hochschule Karlsruhe in 1932 (Trendelenburg 1950). Feldtkeller studied physics in Halle and worked at Siemens Research Laboratories before being appointed in electrical communication engineering at the Technische Hochschule Stuttgart in 1936 (Michel 1994).

20. Stumpf (1908, 226–227). Abraham and von Hornbostel argued that the phonograph, together with the phonometer, could replace the subjective impression of rhythm and pitch with objective and exact measurement methods, heralding a new era in comparative musicology (Abraham and Hornbostel 1904, 226–227).

21. The first volume of his textbook was published in 1923 under the title *Elektronen-Röhren*. Three further volumes followed. The textbook was republished in several languages, including Russian, Bulgarian, and Japanese (Barkhausen 1923; Lunze 1981, 24).

22. Most of the research in Barkhausen's department in Dresden was related to radio tubes and high-frequency technology.

23. Reichardt (1981, 89); Heger (1911). The Sammlung Raumakustik no longer exists, but may have been incorporated into other collections. To set up a "Sammlung und Arbeitsstelle" in order to establish and teach a new field through material practices was rather common at the Technische Hochschule Dresden, which still houses many teaching and research collections in engineering. Barkhausen inherited the Sammlung Telegraphie und Signalwesen (Telegraphy and Signaling Collection) when he was appointed in 1911 (Lunze 1981, 15). The Faculty of Electrical and Computer Engineering at the Technische Universität Dresden now curates the collection, which has since been extended, and makes it accessible to the public.

24. On Pierce's measurements, see E. Thompson (2002, 83–84). Pierce was professor of physics at Harvard University and worked on wireless and electric circuits. Most likely, Heger did not know about Pierce and his use of the telephone receiver and galvanometer. Pierce encountered problems with telephone receivers similar to those that Heger had with phonograph membranes. Pierce's telephone receiver was just not sensitive enough. To make it work, he had to tune the receiver to just a single frequency, and even then, it indicated sound intensity only qualitatively, not quantitatively.

25. See *Personalverzeichnis* (1922, 38).

26. See Meyer (1934, 134–140; 1939, 12) for the unit phon.

27. See Barkhausen and Lewicki (1924); Barkhausen and Tischner (1927). Horst Tischner was Barkhausen's assistant from 1925 to 1927. He completed his doctoral dissertation on sound propagation in tubes and sound absorption in Dresden in 1929 or 1930 while working on sound motion pictures at the AEG Research Institute in Berlin, which he had joined in 1927. After World War II, Tischner became professor of technical physics at Tübingen and finally professor for theoretical electrical engineering at the Technische Hochschule Hannover (Braunschweigische Wissenschaftliche Gesellschaft 1969).

28. This arrangement, producing test sounds by means of a buzzer that were transmitted to the operator by telephone, is reminiscent of the equipment that Emily Mary Smith and Frederic Charles Bartlett had used to test British hydrophone operators during World War I (Smith and Bartlett 1919, 103–106; see also section 3.5 in this volume).

29. Georg Oskar Schubert, for example, completed his dissertation with a systematic study of microphones in 1925 (G. O. Schubert 1927). Schubert went to Siemens Research Laboratories and started to work on television in 1930 (Felgel-Farnholz 1983). Ernst Hormann completed his thesis on magnetic sound recording in 1932, while Rudolf Otto handed in a thesis on the noise produced by the contacts of carbon microphones in 1934 (Hormann 1932; Otto 1935).

30. Kluge (1933); Költzsch (2003, 14). For acoustic filters, see G. W. Stewart (1921, 1922). Stewart referred to George W. Pierce with regard to electric filters. Karl Willy Wagner had worked on the theory of electric filters during World War I as well (Wagner 1919).

31. In the list of members of the Helmholtz Society in May 1937, for example, Barkhausen was listed as a representative for technical physics along with Erwin Meyer, Erich Waetzmann, and Hermann Backhaus (Deutsches Museum Archiv, NL 053, Zenneck 0627).

32. Knoblauch (1941–1942, 25); Lunze (1981, 20); Wein (2011, 51–54). Hilde's daughter Marlene Barkhausen studied technical physics as well. She completed her

diploma thesis with her father in 1936 and went on to work for Siemens and Halske (Wein 2011, 53; Lunze 1981, 121).

33. Zenneck (1899b, 1904). See also Yeang (2012, 203–207) for Zenneck's use of the cathode ray oscilloscope in ionospheric research.

34. Zenneck (1914); Rukop and Zenneck (1914). After being Zenneck's assistant in Danzig and Munich, Rukop joined Telefunken, where he became head of research in 1925.

35. Angerer and Ladenburg (1921, 319) and section 3.2.1 in this volume.

36. Zenneck made quite derogatory remarks about the hybrid curriculum that placed technical physicists between science and engineering, calling its graduates "amphibians" in a letter to Otto Petersen. See Zenneck to O. Petersen, 16 January 1926, and Zenneck to Erich Waetzmann, 27 October 1927, Deutsches Museum Archiv, NL 053, Zenneck 1061 and 1162. The separation of the curricula meant that Knoblauch and Zenneck kept out of each other's way when it came to divergent ideas about the training of technical physicists.

37. "Diplom-Prüfungsordnung für Studierende der Technischen Physik," 13 March 1924, and "Anzahl der Studierenden der technischen Physik an der Technischen Hochschule München," Deutsches Museum Archiv, NL 053, Zenneck 0098 and 0099. Slightly revised regulations were passed in 1929, keeping the distinction between Track A and Track B in place.

38. See list of numbers of graduates from technical physics at the Technische Hochschule München, 1930s. Deutsches Museum Archiv, NL 053, Zenneck 0075.

39. Mauz to Jacoby, 26 October 1931, and Mauz to Zenneck, 28 December 1931, Deutsches Museum Archiv, NL 053, Zenneck 0750.

40. "Zenneck, Jonathan Adolf Wilhelm," 4 January 1952, Deutsches Museum Archiv, NL 053, Zenneck 0062.

41. In a letter to the Deutsches Museum of 16 July 1928, regarding the improvement of the acoustics in its Hall of Fame, Zenneck mentioned that he and his students had already been working in the field of architectural acoustics for several semesters (Deutsches Museum Archiv, NL 053, Zenneck 0806).

42. Reiher published several related papers, including one on architectural acoustics (Reiher 1927). Among Zenneck's archived documents is a list of publications originating in the Laboratory for Technical Physics between 1927 and 1931, specifying a number of publications on sound and sound insulation (Deutsches Museum Archiv, NL 053, Zenneck 0104). Nonetheless, Zenneck never mentioned Reiher or cited him in a publication. All this suggests that he regarded Reiher and others at Knoblauch's laboratory as engineers and not scientists, and thus ignored their work. To my knowledge, Reiher did not cite Zenneck or his students either.

43. Zenneck, "Zur Raumakustik," undated, Deutsches Museum Archiv, NL 053, Zenneck 0728.

44. Strutt (1934, 463–480). See also Berg and Holtsmark (1935a); Stumpp (1936). M. Strutt discussed oscillograms and the work of Zenneck and his students in his chapter on architectural acoustics in the *Technische Akustik* handbook.

45. The American doctoral student Wilmer L. Barrow worked on model experiments in architectural acoustics in 1929 and completed his dissertation in 1930 on a buzzer that produced a wail sound for acoustic testing (Barrow 1931). Barrow came with a recommendation from MIT electrical engineer Ernst A. Guillemin; Guillemin to Zenneck, 22 April 1929. In a letter to the administration of the Technische Hochschule, Zenneck asked to extend Barrow's studies. The copy of the letter refers to the winter semester of 1929–30 (both Deutsches Museum Archiv, NL 053, Zenneck 0082). Barrow later became professor of electrical engineering at MIT and a member of the MIT Radiation Laboratory during World War II.

46. Hans Hoffmann, 17 April 1935; Richard Richter, 17 April 1935; and Hermann Erdlen, 25 April 1935, all Deutsches Museum Archiv, NL 053, Zenneck 0301.

47. Zenneck to Hochbauabteilung, 15 May 1935, Deutsches Museum Archiv, NL 053, Zenneck 0300. The disagreement between Zenneck's scientific expertise and the musician's expertise recalls the public debate about the acoustics of the Boston Symphony Hall. Wallace Sabine had studied the acoustic properties of the hall and consulted on its planning in the early 1900s. Musicians and critics, however, did not consider the resulting acoustics to be favorable (E. Thompson 2002, 51–57).

48. For the Mannheimer Rosengarten, see Zenneck to Hochbauamt Mannheim, 14 March 1934, Deutsches Museum Archiv, NL 053, Zenneck 0301. For the Theater Hall on the Obersalzberg, see Zenneck to Roderich Fick (Frick), 8 September 1937. For the Bavaria Filmkunst studio, see Zenneck to Lechner, 14 June 1939, both Deutsches Museum Archiv, NL 053, Zenneck 0298. More material on these and other consulting projects can be found in Deutsches Museum Archiv, NL 053, Zenneck 0298, 0300, and 0301.

49. For technical acoustics as a model for the rise of technical physics, see Zenneck, "Die Entwicklung der technischen Physik," c. 1930, 8–10, Deutsches Museum Archiv, NL 053, Zenneck 0728. On the dynamics of radio, sound motion pictures, new construction practices, and acoustics research and development for the case of the acoustical laboratories set up by Johan Holtsmark at the Norges Tekniske Høgskole in Trondheim, Norway, in 1929, see Wittje (2003, 69–156). As well as the market created by industry, students' interest in activities in amateur radio were an important motivation for Holtsmark's involvement in acoustical research.

50. On Zenneck's correspondence with industry to place students with a background in acoustics, see Atlas Werke to Zenneck, looking for one or two men for underwater sound, 26 June 1929; Zenneck to Siemens & Halske about several of his

students, 28 January 1935; Atlas Werke to Zenneck, looking for a physicist working on sound technology, 30 January 1935; all Deutsches Museum Archiv, NL 053, Zenneck 0079. See also Robert Bosch AG to Zenneck about Hans Dascher, 19 October 1936, Deutsches Museum Archiv, NL 053, Zenneck 0084. Georg Gehlhoff wrote to Zenneck on 2 December 1930 about industry not taking on enough physicists because of the Great Depression. By 1938, Germany was experiencing a shortage of physicists and engineers; see Zenneck to Jung-Zaeper, 28 September 1938 (both Deutsches Museum Archiv, NL 053, Zenneck 0079).

51. For Trendelenburg and Rukop, see Zenneck to Professor Krüger, Greifswald, 14 May 1930, Deutsches Museum Archiv, NL 053, Zenneck 0990. Zenneck found Trendelenburg even more specialized than Rukop, and doubted that he would be able to lecture on general physics and run a department. For Meyer, see Zenneck to Professor Kirchner, Cologne, 1 February 1937, Deutsches Museum Archiv, NL 053, Zenneck 0970. Zenneck took a positive view of Meyer's contributions to acoustics but argued that he would have to gain more breadth in other fields before he could represent technical physics at the University of Cologne. Meyer was adjunct professor (nichtbeamteter außerordentlicher Professor) at the Technische Hochschule Berlin-Charlottenburg, but at the faculty of machine engineering, not science, and Zenneck assumed that he had only lectured on acoustics there.

52. Barkhausen (1911, 516). As discussed in chapter 2, Henri Poincaré presented a first comprehensive theory of the telephone in 1907 (Poincaré 1907). See also Wagner (1912).

53. Both Barkhausen and Zenneck had industrial experience. Heinrich Barkhausen worked with Siemens and Halske for four years between 1907 and 1911, before he was appointed as professor in Dresden. Jonathan Zenneck worked in the electro-chemical industry (for BASF), again between 1907 and 1911, before joining the Technische Hochschule Danzig.

54. The stories of Conrad Röntgen declining to take out patents for his discovery of X-rays, or of Marie and Pierre Curie doing the same for their process of separating radium, are well known. Wallace Sabine did not take out patents on his reverberation formula either. Though working closely with industry, he was uneasy about the financial aspects of this collaboration. When Jacob Mazer applied for a patent based on Sabine's work in 1911, Sabine and some of his influential friends intervened, going as far as getting the White House and the president of the United States involved (E. Thompson 2002, 71–75, 176–177). Certainly, not all physicists at universities declined to take out patents, but there was a general difference between their attitude toward patents and that of their colleagues at the Technische Hochschulen. See also Wittje (2003, 172, 110n248).

55. See E. Thompson (2002, 99–107) for the careers of Vern Knudsen and Harvey Fletcher as examples of this new generation. In Germany, Heinrich Hecht,

Hugo Lichte, Ferdinand Trendelenburg, and other acousticians never left industrial research.

56. Trendelenburg (1927, 1932, 1935, 1939); Fletcher (1929); Fischer and Lichte (1931). Many of the contributors to Trendelenburg (1927) and Waetzmann (1934a) also came from corporate laboratories.

57. Jonathan Zenneck, for example, saw the United States on the way to taking the lead in experimental physics, as it had in other fields, for example, in electrical engineering (Zenneck 1960, 324, 342; Zenneck to Prof. Henning, PTR, 4 July 1931, Deutsches Museum Archiv, NL 053, Zenneck 0073).

58. See Hagemeyer (1979, 51). For more on the role of the public research institutions, see section 4.4.

59. While the AT&T monopoly in the United States operated around 60 percent of all telephone connections worldwide in the 1920s, Germany had a share of around 9 percent. The UK had around 5 percent, and France around 2.5 percent of all telephone connections. New York had as many telephone connections as the entire UK, and Chicago as many as France. See Hagemeyer (1979, 47, table 1).

60. Hagemeyer (1979, 49, table 2). According to Hagemeyer, Bell Laboratories had a workforce of 589 in its research department, which was not involved in product development, while Siemens Research Laboratories had a workforce of 90.

61. See Fischer (1992) for a social history of telephony in the United States, and Hagemeyer (1979, 53–54) for a comparison between the organization of telephony in the United States and Europe.

62. The Telegraphentechnische Reichsamt was renamed to Reichspostzentralamt (German Central Post Office) in 1928. See Hagemeyer (1979, 320–325). More is said about the Telegraphentechnisches Reichsamt in section 4.4.

63. Many examples of such modular sound systems and instrument racks from the interwar period can be found in the trade catalog collections in the Siemens and Telefunken archives. See, for example, Siemens Corporate Archives 15964. The amplification system of the Deutsches Museum of 1926, which is preserved in the museum's storage rooms, is an early example (Deutsches Museum, inventory number 62457, acquired in October 1926).

64. *Rice Kellogg Verträge*, Telefunken Archiv (2.1.60c), 3417, Historisches Archiv, Deutsches Technikmuseum Berlin. The Rice and Kellogg speaker became the dominant loudspeaker design, especially for radios and smaller systems.

65. In a contract of June 1936, Siemens and Halske, AEG, and Telefunken determined how the different fields of business in electroacoustics should be divided between the two mother companies and Telefunken. The contract was predated to October 1931, almost five years earlier, indicating the time of the original agreement

between Siemens, AEG, and Telefunken. See "Geschäftsvertrag zwischen der Allge-meinen Electricitäts-Gesellschaft zu Berlin, der Siemens & Halske Aktiengesellschaft zu Berlin, der Telefunken Gesellschaft für drahtlose Telegraphie m.b.H. zu Berlin," 18 June 1936, Telefunken Archiv, I 2 060.C 2365.

66. Feldenkirchen (1997, 217–222). On the organization of research and develop-ment at Siemens, AEG, and Telefunken, see Erker (1995, 237–246).

67. Siemens & Halske Akt. Ges., Vakuumverstärkerröhre mit Glühkathode und Hilfselektrode, DRP 300 617, filed 1 June 1916, issued 12 July 1921, DEPATISnet, Deutsches Patent- und Markenamt patent database, https://depatisnet.dpma.de/DepatisNet/. The patent does not carry Schottky's name and does not mention him in the text either.

68. See J. B. Johnson (1925) as an example of this change.

69. For Gerlach and the electroacoustic laboratory, see Heinz Orlich, "Aufzeichnun-gen über seine Tätigkeit im Gerlach—Laboratorium im ZL Siemensstadt 1918–1931," Siemens Corporate Archives 12/LH 866.

70. According to Gerlach's colleague Heinz Orlich, Schottky did not contribute to the idea of the actual speaker and should not have been given credit for its design and patent. Orlich confirmed that Gerlach discussed loudspeaker design with Schottky but argued that Schottky only expressed what "everybody" knew already (ibid.).

71. "Jahresbericht des Zentrallaboratoriums, Geschäftsjahr 1. Oktober 21–30. September 1922," 8, Siemens Corporate Archives LI 869.

72. Schottky (1927). Carl A. Hartmann, who worked with Schottky on the Schottky effect, worked on the testing of microphones around the same time (Hartmann 1922; Trendelenburg 1975, 175, 179).

73. See Trendelenburg (1924); Backhaus and Trendelenburg (1925); Trendelenburg (1925; 1926; 1966, 103–121; 1975, 179–187).

74. Trendelenburg (1975, 177–179); Zenneck (1925); "Lautsprecher zur Eröffung des Deutschen Museums (Angaben von Siemens & Halske)," Deutsches Museum Archiv, NL 053, Zenneck 0291. Trendelenburg reported that the ribbons of the speakers were not durable enough to withstand the strain and had to be replaced several times. According to Zenneck, ten Blatthaller speakers, two ribbon speakers, four ribbon microphones, and two condenser microphones were used at different locations altogether. The amplification system, or at least many parts of it, are in storage at the Deutsches Museum. A large "Giant Blatthaller" weighing 254 kg was registered in 1927 (inventory number 62403). A ribbon microphone registered the same year carries the inventory number 62293. Other objects include the large switchboard (62457), the power amplifier for the Blatthaller (62294), a smaller

Blatthaller (59366), other amplifiers (62281, 62290, 62292, and 62396), and a 64 kg ribbon speaker (59342). The large wooden trumpet seen in figure 4.2 seems to be missing.

75. Trendelenburg (1975, 193–194); Neumann and Trendelenburg (1931). For several reasons, Siemens did not continue the development of the Giant Blatthaller and other very large loudspeakers. One was the variation in the range and loudness of these speakers, which could be huge depending on atmospheric conditions. The Blatthaller was also less efficient than the Rice and Kellogg loudspeaker that Siemens manufactured through patent agreements. The transfer of most electroacoustic activities from Siemens to Telefunken in 1931 in the midst of the Great Depression put an end to work on the Giant Blatthaller and its like.

76. On Lübcke, see Backhaus (1951). Lübcke, Trendelenburg, and Gerdien all filed patents on underwater acoustics for Siemens. See, for example, Lübcke, Wasserschallgerät, DRP 498 637, filed 18 January 1929, issued 8 May 1930; and Trendelenburg, Sende- und Empfangseinrichtung für Unterwasserschallwellen, DRP 481 116, filed 13 August 1926, issued 25 July 1929, DEPATISnet. According to Trendelenburg (1975, 185), Hans Gerdien's patent on an underwater sound transmitter based on magnetostriction was particularly important (Gerdien, Unterwasserschallsender, DRP 449 982, filed 19 January 1927, issued 15 September 1927, DEPATISnet).

77. See, for example, Lübcke, Mit einmaliger Explosion für je eine Echolotung von einem Fahrzeug, insbesondere Luftfahrzeug, aus arbeitender Knallsender, DRP 599 850, filed 21 June 1930, issued 10 July 1934, DEPATISnet; Trendelenburg, Schalltrichter für akustische Richtungsempfangsanlagen, DRP 587 560, filed 27 May 1932, issued 4 November 1933; Gerdien, Richtungshörgerät für Luftschall, DRP 595 120, filed 9 December 1932, issued 29 March 1934, DEPATISnet.

78. See Donhauser (2007) for a history of electroacoustic musical instruments in Germany and Austria in this period, and especially pages 83–99 for the Neo-Bechstein piano. According to Donhauser, up to 150 instruments were built between 1931 and 1933, far fewer than anticipated by the producers. The piano manufacturer Petrof in Prague also produced Neo-Bechstein pianos under license. Oskar Vierling later became Karl Willy Wagner's assistant and worked on several electroacoustic music instruments at the Heinrich Hertz Institute (see section 4.4.1). For the contract between Nernst and Siemens on 28 March 1931 and its transfer to Telefunken on 20 September 1932, see "Patentnutzungsvereinbarung zwischen Siemens und Nernst," Telefunken Archiv, I 2 600 3504.

79. AEG presented its own three-volume history of its research and development in 1965 (Schweder 1965). In the chapter on physics technology and research, Carl Ramsauer argued that AEG was rather passive toward physical research until it founded its research institute in 1928 (Ramsauer 1965). The low research activity at AEG compared to Siemens was probably also related to the circumstance that AEG

was mainly active in the field of high-voltage engineering. As Trendelenburg pointed out, Siemens had carried out little research in high-voltage engineering in the early 1920s (Trendelenburg 1975, 46).

80. See Ramsauer (1965, 401). Klingenberg was an electrical engineer and a board member of AEG.

81. See Brüche (1938); Lorenz (2004); Schweder (1965); Erker (1995, 237–239); *Jahrbuch des Forschungs-Instituts der Allgemeinen Elektricitäts-Gesellschaft* (1930–1938). This scattering, which can only be explained by quantum mechanical wave theory, is today known as the Ramsauer-Townsend effect. The Technische Hochschule Berlin-Charlottenburg appointed Ramsauer as an honorary professor in 1931.

82. *Jahrbuch des Forschungs-Instituts* (1930 1:4). According to Lorenz (2004, 11), the AEG Research Institute started with twenty-three scientists in 1928. In 1929, it employed forty-five scientists and engineers, along with fifteen auxiliary workers and eighty mechanics in the workshop.

83. Lichte (1965, 425). See Mühl-Benninghaus (1999, 119–121, 138–139) on the technical problems experienced by the UFA film company with this early recording and reproduction apparatus.

84. *Jahrbuch des Forschungs-Instituts* (1930, 1:5).

85. E. Thompson (2002, 246–247); Jossé (1984, 238–242). It was *The Singing Fool*, released in 1928, that finally conquered international markets (Mühl-Benninghaus 1999, 116, 134).

86. My translation. "Seit Jahrzehnten hat sich Erfindergeist mit der Aufgabe des sprechenden Films befaßt, aber erst die Entwicklung der technischen Hilfsmittel auf mannigfachen Gebieten hat das Problem des Tonfilms seiner Lösung und praktischen Auswertung zugeführt. Die fortschreitende Entwicklung auf dem Radiogebiet hat dem Tonfilmproblem ebensoviel weitergeholfen wie die Vervollkommnung der Elektroakustik auf dem Gebiete der Sprechmaschinen und Schallplatten und die Verbesserung in der Herstellung des Films durch die chemische Industrie."

87. See *Jahrbuch des Forschungs-Instituts* (1930, 1:13); Fischer and Lichte (1934, 349–350); Ruhmer (1901). Of course, Lichte's reference to Simon and the electric arc created a link to his own vita. See Jossé (1984, 106–122) for other entrepreneurs aiming to develop a sound-on-film system for sound motion pictures before World War I.

88. See Engl (1927) for the Tri-Ergon system. Both Vogt and Massolle were wireless telegraphers during World War I. The failure of the Tri-Ergon consortium to develop a commercially successful system may be explained by its small scale and the lack of unrestricted backing from a larger company (Jossé 1984, 140–208; Mühl-Benninghaus 1999, 21–41).

89. For the transition to sound motion pictures in the United States, see Crafton (1997); Eyman (1997). See E. Thompson (1997) for the role of acoustics research, especially architectural acoustics, in that transition.

90. On Fischer, see Borgnis (1961). Fischer became professor of technical physics at the Eidgenössische Technische Hochschule (ETH), Zürich, in 1933.

91. Other members of the AEG team were Friedrich Wilhelm Hehlgans, who had written a dissertation on piezo-acoustic transducers at the University of Jena, and Friedrich Wilhelm Dustmann, later the sound engineer for many early German sound motion pictures. See Wulff and Schumacher (2001, 1202) for the early development of sound motion pictures at AEG and Siemens. On Narath, see M. Engel (1997). See also Hehlgans (1928). According to Lorenz, the AEG acoustics laboratory had a staff of twenty-five in 1931, making it the largest in the AEG Research Institute, which had a total workforce of seventy-nine at the time (Lorenz 2004, 12).

92. See *Jahrbuch des Forschungs-Instituts* (1930, 1:11–25).

93. See Jossé (1984) and Mühl-Benninghaus (1999) on the beginnings of sound motion pictures in Germany and the various agreements between German actors to prevent a takeover by the American film industry.

94. See Fischer and Lichte (1931, 1934). For Klangfilm's marketing of the different systems, see, for example, the brochure "Typen der Tonfilm-Aufnahme - Apparate der Tobis-Klangfilm," c. 1931, Siemens Corporate Archives, file 8353. Klangfilm sold recording systems for sound-on-film only, for needle systems only, and combined systems that could record sound in both ways.

95. *Jahrbuch des Forschungs-Instituts* (1930, 1:27–61; 1931, 2:43–65).

96. Mielert (1985) and "Telefunken—Organisation," 2 March 1939, Siemens Corporate Archives 54 Li 63, Telefunken. Albert Narath also joined Telefunken after 1931.

97. Rössler (2006, 39, 43, 56). Stenzel also wrote the chapter on loudspeakers in the *Technische Akustik* handbook of 1934 (Waetzmann 1934a, 254–300).

98. For Stenzel's references to underwater acoustics and submarine communication and detection, see *Jahrbuch des Forschungs-Instituts* (1930, 1:25, 43, 44). On the patent, see Arrangement for directional transmission and reception with a plurality of oscillators, United States Patent Office 1,893,741; application filed 9 January 1929 by Heinrich Hecht and Heinrich Stenzel for Electroacoustic GmbH. The German patent application was filed a year earlier, on 14 January 1928, DEPATIS-net. For Stenzel's publications on echo sounding, see Stenzel (1926). Stenzel probably completed a dissertation in mathematics with Edmund Landau in 1920 (Stenzel 1922). While I have no evidence that this is the same Heinrich Stenzel, it seems very likely given the mathematical character of most of his publications in acoustics.

99. See Cahan (1989) and chapter 2 in this volume on the Physikalisch Technische Reichsanstalt; Rasch (2006) and chapter 3 on the Kaiser Wilhelm Foundation.

100. "Kuratorium der Physikalisch-Technischen Reichsanstalt, Sitzung am 14. und 15. März 1928," pp. 13–14, Deutsches Museum Archiv, NL 053, Zenneck 0683. The only activity of the Reichsanstalt that fell in the field of acoustics was the testing of tuning forks, which was connected to precision mechanics.

101. On the trautonium, see Schenk (1997); Stange (1989, 85–101); Donhauser (2007, esp. 65–78, 131–143). Telefunken started to produce the trautonium for home use in 1933, which became known as the Volkstrautonium. Like the Neo-Bechstein, this early electroacoustic musical instrument did not become a sales success. Largely because of Paul Hindemith's compositions and Oskar Sala's continued development and promotion of the instrument after World War II, the trautonium has not been forgotten, unlike other early electroacoustic musical instruments.

102. See Lübcke (1943) and Wagner's detailed scientific autobiography of 1936, *Laufbahn und wissenschaftliche Arbeiten*, Nachlass Karl Willy Wagner, Archiv der Berlin-Brandenburgischen Akademie der Wissenschaften, Berlin.

103. This journal seems to have been central to establishing the notion of *Nachrichtentechnik* as electrical communication engineering, previously not a common concept, in Germany. *Elektrische Nachrichten-Technik* also published many papers on electroacoustics.

104. Sitzungen des Telegraphentechnischen Reichsamtes, 14–17 April 1926, 20–28 April 1927, 18–21 April 1928, 17–20 April 1929, Deutsches Museum Archiv, NL 053, Zenneck 0204 and 0680. Presentations in acoustics included Barkhausen on new acoustic investigations in 1927, Zenneck on architectural acoustics in 1928 and 1929, and Meyer on gramophone records as a tool in acoustic measurement in 1929.

105. See the annual reports of the Heinrich Hertz Institute and Wagner, "Bericht über meine seit dem 31. 1. 1936 schwebende Angelegenheit," 27 February 1940, p. 2, Nachlass Karl Willy Wagner. The head of the department of acoustics, Erwin Meyer, was not appointed adjunct professor until 1934, despite having finished his habilitation in 1928. Both Wagner and Meyer indicated tensions between the Technische Hochschule and the Heinrich Hertz Institute. See, for example, Meyer in *Verleihung der Ehrendoktorwürde* (1958, 14).

106. Wagner, "Bericht über meine seit dem 31. 1. 1936 schwebende Angelegenheit," 27 February 1940, p. 2, Nachlass Karl Willy Wagner. In Hagemeyer's comparative study, he argues that Schwingungsforschung was a specifically German approach, distinct from U.S. approaches, which treated the different types of oscillations separately (Hagemeyer 1979, 54 and 70). As discussed earlier, the research program of Schwingungsforschung was not unique to Wagner but had already

been formulated by Barkhausen in 1906, before Wagner joined Simon's institute (Barkhausen 1907).

107. There were, however, some important differences. Trendelenburg came from an influential upper-middle-class family, Meyer from a less affluent background. Trendelenburg worked for Siemens throughout his career, while Meyer worked only in public institutions.

108. Meyer compared these forces to Vilhelm Bjerknes's hydromechanical forces (Meyer 1923). After the war, this approach to acoustics was old-fashioned and removed from an electroacoustic understanding of sound. As Guicking notes, Meyer worked as Otto Lummer's assistant from 1923 to 1924 and impressed Hans Salinger with his homemade radio apparatus, which suggests that he was a skilled radio amateur. Salinger, at that point a Reichspost researcher, recommended Meyer to Wagner (Guicking 2012, 4–5).

109. Erwin Meyer, Einrichtung zur Klanganalyse, DRP, Patentschrift 511146, Klasse 42g, Gruppe 1/01, filed on 12 October 1928, issued on 16 October 1930, DEPATISnet. It is probably not surprising that Trendelenburg worked on a similar project at the same time at Siemens.

110. On Meyer, see Kuttruff (1969); *Verleihung der Ehrendoktorwürde* (1958); Schroeder (1972, 2000); Cremer (1972); Guicking (2012). Meyer's lectures on electroacoustics, held at the Institution of Electrical Engineering in London in 1937, were published as a textbook in 1939 (Meyer 1939).

111. Heinrich-Hertz-Institut für Schwingungsforschung (1933); Lübcke (1943, 80). Wagner became head of the Noise Abatement Committee of the German engineering association Verein Deutscher Ingenieure in 1930. On Wagner's work on noise abatement, see Wagner (1931, 1936a); Bijsterveld (2008, 108–109); on his vocal synthesizer, see Wagner (1936b).

112. See "Gründung einer 'Gesellschaft für elektrische Musik'" (1932).

113. See Donhauser (2007, 111–117, 133–139); Fuchs (1933, 211–212).

114. See Donhauser (2007); Vierling (1936); Voigt (1988). The only remaining Elektrochord is on display at the Deutsches Museum in Munich.

115. Institut für Schwingungsforschung der Technischen Hochschule Berlin (1938, 61). Staff numbers for the earlier years are not available.

116. Knoblauch (1941/42, 68); Verein zur Förderung des Institutes für Schall- und Wärmeforschung, "Haupt-Daten über die Vorgeschichte der Gründung des Instituts für Schall- und Wärmeforschung der Technischen Hochschule Stuttgart," 15 October 1932, 57, Karton 149, Personalakten Hermann Reiher (Prof) 006445, Universitätsarchiv Stuttgart.

117. Verein zur Förderung des Institutes für Schall- und Wärmeforschung, "Haupt-Daten über die Vorgeschichte der Gründung des Instituts für Schall- und Wärmeforschung der Technischen Hochschule Stuttgart." In the early documents, the institute is sometimes called Anstalt für Schall- und Wärmetechnik or Institut für Schall- und Wärmetechnik. The name Institut für Schall- und Wärmeforschung was finally chosen.

118. As a provider of acoustic knowledge and measurement for local and regional markets, Reiher's institute was similar to other acoustics laboratories, for example, the one that Johan Peter Holtsmark established at the Norges Tekniske Høgskole in Trondheim around 1929 (Wittje 2003, esp. 107–108).

119. See *Programm der Württembergischen* (1929, 38).

120. Erich Regener to Richard Grammel, Rector of the Technische Hochschule Stuttgart, 19 April 1948, 57, Karton 149, Personalakten Hermann Reiher (Prof) 006445, Universitätsarchiv Stuttgart. Regener also cited evaluations of Reiher by Erwin Meyer and others. Regener's very negative characterization of Reiher in 1948 was clearly reinforced by the events during the Third Reich, when Regener was forced into provisional retirement and Reiher was appointed professor of technical physics by the National Socialist regime. Clearly, however, Regener had already opposed Reiher's appointment as professor in the early 1930s, and Reiher was only appointed after Regener lost his power.

121. See *Programm der Württembergischen* (1933, 55).

122. As I argue above, the communities of physics and electrical engineering were closely connected in the interwar period, and actors from both sides were easily able cross the disciplinary boundaries. The same was not true between physics and civil engineering.

123. Ferdinand Trendelenburg, the main acoustician at Siemens Research Laboratories, was honorary professor at the Friedrich Wilhelm University; Carl Ramsauer, head of AEG Research Institute, was honorary professor at the Technical University. Karl Willy Wagner, the Heinrich Hertz Institute's director, and the directors of its divisions, were all adjunct professors at the Technische Hochschule Berlin-Charlottenburg.

124. The Deutsche Forschungsgemeinschaft initiated the *Akustische Zeitschrift* "at the suggestion of Prof. Dr. Stark" (Grützmacher and Meyer 1936). Stark was president of the Forschungsgemeinschaft from 1934 to 1936, a position from which he ruled according to the *Führerprinzip* (leader principle), the foundation of Nazi political authority.

Chapter 5

1. See Parey (1935). Electrical engineer Walter Parey was editor of the *Zeitschrift des Vereins Deutscher Ingenieure*.

2. My translation. "Das Zeitalter der 'reinen Vernunft', der 'voraussetzungslosen' und 'wertfreien' Wissenschaft ist beendet" (Krieck 1936, 1).

3. My translation. "Im neuen Deutschland ist die Rundfunktechnik nicht eine Wissenschaft, die einigen Eingeweihten Unterlagen gibt, ihre Kenntnisse zu erweitern und diese an einem technischen Mittel auszutoben. Im Gegenteil: Seit der Machtübernahme durch die nationalsozialistische Regierung wurde der deutsche Rundfunk hineingetragen ins Volk. Er wurde der Mittler zwischen Staatsführung und Bevölkerung. Er wurde der direkte Erlebnisgestalter der großen Feiern des deutschen Volkes" (Reprint from *NS-Funk*, 10 July 1938, in *Großübertragungsanlagen im Dienste der Volksführung* [Large amplification systems in the service of the people's leadership], Telefunken electroacoustics pamphlet 8/38, Siemens Corporate Archives, file 8014).

4. See W. Zellner, VDI Berlin, "Die Lärmbekämpfung, eine Aufgabe des ganzen Volkes—Die Reichswoche ohne Lärm vom 7. Bis 13. April 1935," Mitteilung der Pressestelle RTA 32, Reichsgemeinschaft der technisch-wissenschaftlichen Arbeit, Deutsches Museum Archiv, NL 053, Zenneck 0307.

5. Erich Regener to Richard Grammel, rector of the Technische Hochschule Stuttgart, 19 April 1948, 57, Karton 149, Personalakten Hermann Reiher [Prof] 006445, Universitätsarchiv Stuttgart. As I argue in the chapter 4, Regener's negative characterization of Reiher in 1948 was clearly reinforced by the events during the Third Reich, but he clearly had already opposed Reiher's appointment as professor in the early 1930s.

6. See Hentschel (2007) for the mentality and general attitude of German physicists in the early postwar period regarding the Third Reich and their own conduct. Wagner's membership meant that he was registered as a political follower of the NSDAP in 1946. He appealed against the registration since he saw himself as a victim rather than a supporter of the Nazi establishment. See Wagner to Spruchkammer Bad Homburg, 27 April 1946, Nachlass Karl Willy Wagner, Archiv der Berlin-Brandenburgischen Akademie der Wissenschaften, Berlin.

7. See, for example, Oskar Hecker, Reichsanstalt für Erdbebenforschung, to Zenneck, 13 October 1928, and Zenneck to Hecker, 16 October 1928, both in Deutsches Museum Archiv, NL 053, Zenneck 0944. Zenneck opposed Hecker's proposal to nominate Wagner for the Abbe prize, stating that Wagner was, without doubt, a capable man, but more so a capable businessman, and that his nomination would meet strong criticism. In 1931, Wagner tried to convince Zenneck to participate in the meeting of the International Union of Radio Science in Copenhagen, after the

boycott against German scientists was lifted. Zenneck was, again, fiercely opposed against a German participation, arguing that self-respect forbade German scientists from participating after the boycott. Zenneck, who belonged to the right-wing nationalist political spectrum, most likely opposed a renewed German participation in the international unions even before he was contacted by Wagner. But that Wagner proposed a German "delegation under the leadership of the Heinrich Hertz Society" certainly did not make Zenneck more sympathetic to the idea (Wagner to Zenneck, 22 May 1931, and Zenneck to Wagner, 27 May 1931, both Deutsches Museum Archiv, NL 053, Zenneck 1166).

8. Wagner to Spruchkammer Bad Homburg, 27 April 1946, and Wagner An den Herrn Direktor der US-Militärregierung in Bayern, 13 February 1947, both Nachlass Karl Willy Wagner. See also Gundlach (1978, 6) about the changes at the Heinrich Hertz Institute in 1936. In 1943, however, Ernst Lübcke published an article about Wagner's contributions to acoustics research in the *Akustische Zeitschrift*, on the occasion of Wagner's sixtieth birthday (Lübcke 1943). An indication of Wagner's rehabilitation within the NS establishment around this time was his activity as a scientific adviser to the German Navy from 1943 to 1945.

9. See VDE, "Einheit der Frequenz 'Helmholtz,'" 8 November 1939, Deutsches Museum Archiv, NL 1186, Zenneck 028.

10. König (2004). Other peoples' products were the *Volkskühlschrank* (peoples' refrigerator), the *Volksfernseher* (peoples' television), and the *Volkswohnung* (people's housing). Only the *Volksempfänger* was mass produced before Hitler started the war, which set an end to the plans of an NS consumer society.

11. König (2003a, 91–93). According to Wolfgang König, the main reason that large sections of the working class could not afford a radio was not the price of the receiver but the monthly compulsory fees, which funded Goebbels's Ministry of Propaganda.

12. Ibid., 100. The propaganda department of the NSDAP had plans to install a total of 6,600 loudspeaker poles in all German cities and towns with more than 12,000 inhabitants. For smaller towns and villages, the propaganda department developed plans for the *Gemeinderundfunk* (township radio broadcasting).

13. "Die größte Lautsprecheranlage der Welt" (1933) and "Die Telefunken-Großlautsprecheranlage auf dem Tempelhofer Feld" (1933). According to both articles, this was the largest sound amplification system worldwide.

14. The meeting place at the Tempelhofer Feld, for example, was 800 meters wide and 1 kilometer deep. If broadcast from one single location, the sound would take up to three seconds to reach some members of the audience.

15. See *Materialiensammlung über Lautsprecher*, Telefunken Archiv (1.2.60c), 972, especially the pamphlets *Nachrichten aus der Elektroakustik*, Folge 3 and 4, 1935.

16. Regarding the continuity of military research in acoustics in Britain, see Hackmann (1984, 97–231), for underwater acoustics and the development of asdic (today
known as sonar), and Zimmerman (2001), for Tucker's acoustic mirrors and the
development of radar for air defense.

17. Sitzungen des Telegraphentechnischen Reichsamtes, 14–17 April 1926, 20–28
April 1927, 18–21 April 1928, 17–20 April 1929, Deutsches Museum Archiv, NL 053,
Zenneck 0204 and 0680).

18. The range of sound telegraphy apparatus was tested with two Turkish subma-
rines in the Sea of Marmara in December 1931 and January 1932, and with two
Dutch submarines in Den Helder in March 1932. See "Bericht über die Abnahmever-
suche im Marmarmeer mit 2 türkischen Unterseebooten und dem Linienschiff
'Jawus Selim' am 16./17. Dezember 1931 und 3. Januar 1932," 30 March 1932, and
"Bericht über die Abnahmeversuche zwischen den holländischen Unterseebooten
0.14 und 0.15 am 14. März 1932 in Den Helder," 30 March 1932, both in Deutsches
Museum Archiv, NL 053, Zenneck 0293.

19. See Backhaus (1951) for Lübcke. All three, Lübcke, Trendelenburg, and Gerdien
took patents on underwater acoustics for Siemens. See, for example, Lübcke, Wasserschallgerät, DRP 498 637, filed on 18 January 1929, issued 8 May 1930; and Trendelenburg, Sende- und Empfangseinrichtung für Unterwasserschallwellen, DRP 481
116, filed on 13 August 1926, issued 25 July 1929, DEPATISnet, Deutsches Patent-
und Markenamt patent database, https://depatisnet.dpma.de/DepatisNet/. According to Trendelenburg (1975, 185), Hans Gerdien's patent on an underwater sound
transmitter based on magnetostriction was specifically important (Gerdien, Unterwasserschallsender, DRP 449 982, filed on 19 January 1927, issued on 15 September
1927, DEPATISnet).

20. See, for example, Lübcke, Mit einmaliger Explosion für je eine Echolotung von
einem Fahrzeug, insbesondere Luftfahrzeug, aus arbeitender Knallsender, DRP 599
850, filed on 21 June 1930, issued on 10 July 1934; Trendelenburg, Schalltrichter für
akustische Richtungsempfangsanlagen, DRP 587 560, filed on 27 May 1932, issued
on 4 November 1933; Gerdien, Richtungshörgerät für Luftschall, DRP 595 120, filed
on 9 December 1932, issued on 29 March 1934, DEPATISnet.

21. See Lilienthal-Gesellschaft für Luftfahrtforschung (1937) and Deutsche Akademie der Luftfahrtforschung (1939).

22. See Zimmerman (2001) for the case of Britain. Acoustic location was, however,
still used for directing searchlights for anti-aircraft artillery during World War II.

23. See Institut für Schwingungsforschung der Technischen Hochschule Berlin
(1938, 4, 31).

24. See Institut für Schwingungsforschung der Technischen Hochschule Berlin
(1940, 3–4).

25. See Rössler (2006, 127–141) for the development and testing of Alberich and Fafnir, and Guicking (2012, 12–13) and Schroeder (2000, 181) about Meyer.

Chapter 6

1. My translation. "Auf dem Gesamtgebiet der Akustik sind in den letzten Jahren außergewöhnlich große Fortschritte erzielt worden, wobei der Anstoß weitgehend durch rein technische Probleme gegeben wurde. Es liegt hier ein Musterbeispiel dafür vor, wie eng reine und angewandte Physik heute miteinander verbunden sind und wie stark sie sich gegenseitig beeinflussen und befruchten. Die reine Akustik ist in erster Linie durch technische Fragestellungen aus ihrem Dornröschenschlaf erweckt worden, und die technische Akustik verdankt ihren Aufschwung zum großen Teil der Tatsache, daß sie auch alle Errungenschaften der reinen Physik in ihre Dienste gestellt hat" (Waetzmann 1934a, vol. 1, v, preface).

2. Despite criticism by historians of science in recent decades of the concepts of "modern physics" versus "classical physics," as of concepts of modernity in general, scientists and historians have followed the distinctions of classical and modern physics at least since the 1930s. In this distinction, acoustics clearly features as classical physics.

3. See also J. Hughes (1993, chapter 4), and Wittje (2003, 159).

4. Schrödinger pointed out the limits of analogy, however, between what he already called classical wave mechanics, including acoustics, and the new quantum mechanics (Schrödinger 1926, esp. 114). During World War I, Schrödinger was an artillery officer of the Austrian Army and taught meteorology to artillery personnel. He published a paper on sound propagation in the free atmosphere in 1917, which was directly relevant for artillery ranging. It remained his only publication in the field of acoustics (Schrödinger 1917).

5. Landé (1930, 17–18). Just for the record, Landé was involved in sound ranging during World War I as well. He worked at the APK in Berlin, with Born and Ladenburg (see the transcript of an oral history interview with Dr. Alfred Landé by Thomas S. Kuhn and John Heilbron in Berkeley, California, March 5, 1962, Niels Bohr Library, Center for History of Physics, American Institute of Physics, College Park, MD).

6. Stewart cited Landé as the one who made the suggestion to him in the first place, implying that the two had met during Landé's lecture series (G. W. Stewart 1931, 327).

7. To my knowledge, Hecht's booklet was a rather isolated approach that did not find any resonance or provoke any reaction in the physics community. Heinrich Hecht was not the only acoustician to reject the theory of relativity. Dayton Clarence Miller, president of the Acoustical Society of America from 1931 to 1933,

claimed to have disproved Einstein's theory (E. Thompson 2002, 105). Hecht referred to Miller's experiments (Hecht 1954, 30).

8. See Ramsauer (1949, 1–7). Ramsauer presented the same view regarding the physics-technology relationship in several other papers published in the 1949 book.

9. Ramsauer (1949, 3). I would like to thank Klaus Hentschel for drawing my attention to Ramsauer's modification of the figure after the war. See also Schmithals (1980).

10. In a reply in 1948, Richard Vieweg criticized the diagram that Ramsauer used to visualize the relationship between physics and technology (see figure 6.1). According to Vieweg, it was known as the "Ramsauersche Molluske" (Ramsauer's mollusk; Vieweg 1948, 16). Similar to Waetzmann in 1934 and others, Vieweg pointed out the strong mutual relationship between science and technology, where the flow of inspiration could go both ways. While Vieweg agreed that chemistry had received new impulses by the advances in atomic physics, he argued that the nature of chemistry was distinct from physics.

11. See Rammer (2004, 561–564) about Meyer's appointment at Göttingen. In Göttingen, Meyer carried on with his research on sound absorption in liquids, related to submarine warfare, for the British Department of Scientific and Industrial Research.

12. See *Verleihung der Ehrendoktorwürde* (1958); Schroeder (1972, 2000); and Cremer (1972).

13. See especially Kurz, Parlitz, and Kaatze (2007). In many ways, the research program on Schwingungsforschung originated at the Institut für Angewandte Elektrizität of the University of Göttingen prior to World War I, where Heinrich Barkhausen formulated it in his doctoral dissertation in 1907, and where Karl Willy Wagner obtained his doctorate in 1910. The Drittes Physikalisches Institut was actually created by a fusion of the former institutes of Angewandte Elektrizität and Angewandte Mechanik (Applied Mechanics) in 1947.

14. Charles Withers, in his work on *Geography, Science and National Identity*, wants us to "consider how given forms of geographical knowledge themselves came to constitute the idea of Scotland as a national space" instead of presuming national styles and movements in a national context (Withers 2001, 15).

15. See Erker (1995) and Boersma (2002). For the international business history of the Siemens Company, see Feldenkirchen (1997).

16. Institutions like the Technische Hochschule of Vienna and the Eidgenössische Technische Hochschule in Zurich were certainly not copies of a "German model of the Technische Hochschule" but themselves actors in defining what a Technische Hochschule should be like.

17. See, for example, Tekniska fysikers förening (1982) for the establishment of technical physics at the Kungliga Tekniska Högskolan in Stockholm, Sweden, during the interwar period, and Wittje (2003, 60–62 and 102–104) for the Norges Tekniske Høgskole in Trondheim, Norway.

18. For Aigner, see Kurzel-Runtscheiner (1953). For Osswald and acoustic research at the ETH Zurich, see Eggenschwiler and Fischer (2014). For Fokker and especially his work on the theory of music, see Hiebert (2014, 197–217). See also Bijsterveld (2008, 124) and the transcript of the oral history interview with Adriaan Fokker by John Heilbron in Beekbergen, Holland, 1 April 1963, Niels Bohr Library, Center for History of Physics, American Institute of Physics, College Park, MD. For Holtsmark and his agenda in technical acoustics, see Wittje (2003, esp. 69–156).

19. See, for example, the chapters on sound motion pictures and magnetic recording in Waetzmann (1934a).

20. I avoid using the term "electronics" for the interwar period because it did not come into general use for radio and amplifier technology until World War II.

21. Fletcher (1923); Barkhausen and Lewicki (1924). For his dissertation, Fletcher worked with Robert Millikan on the oil drop experiment to measure the charge of the electron. Fletcher started his career with Bell during World War I, building an underwater sound detector.

22. Shot noise defined a lower limit of noise that could theoretically be achieved in tube amplifiers. For practical matters of sound amplification, however, other sources of distortions dominated. In his widely used handbook on amplifier tubes, Heinrich Barkhausen discussed the limits of tube amplification through distortions that were partly caused by the tube itself, partly by other components of the electric circuit, but did not mention shot noise (Barkhausen 1923, 1, 109).

23. My translation. "Lärm wird jeder unerwünschte Hörschall genannt, gleichgültig, ob er ein Ton oder ein Geräusch ist. Die Begriffe Lärm und Geräusch sind nicht gleichbedeutend" (Berger 1934, 1).

24. In the volume on *Akustik* in the *Handbuch der Physik* of 1927, noise as a nuisance (Lärm or Störschalle) did not feature. In the two volumes of 1934's *Technische Akustik* in the *Handbuch der Experimentalphysik*, noise abatement had gotten its own eighty-page chapter. Berger was, in fact, one of the first acousticians to work on noise abatement in Germany, and he completed his dissertation on sound transmission through walls at the Technische Hochschule Munich in 1911. During World War I, Berger worked on sound ranging (Berger 1926, esp. 56–57).

25. My translation. "Oft wird 'Lärm' gleich 'Geräusch' gesetzt. Das sind aber zwei verschiedene Dinge. Der Ausschuß für Einheiten und Formelgrößen (AEF) definiert (Entwurf Nr. 37) folgendermaßen [*Elektrotechnische Zeitschrift* 54 (1933) S. 783]:

'Geräusch': Schallschwingungen, die sich aus einem kontinuierlichen Tonspektrum oder aus einem solchen mit sehr vielen Einzeltönen beliebiger Höhe zusammensetzt.

'Lärm': Jede Art von Schallschwingungen, die eine gewollte Schallaufnahme oder die Stille stört.

Das Geräusch ist hiernach eine rein physikalische Größe; Die Messung von Geräuschen kann also, mindestens im Prinzip, keine Schwierigkeiten bieten. Der Lärmbegriff enthält dagegen neben den physikalischen auch psychologische Elemente; diese lassen sich mit physikalischen Messungen nicht oder bestenfalls nur mittelbar erfassen."

26. My translation. "So ist beispielsweise der Klang eines Klaviers nicht streng periodisch, im Moment des Anschlags ist dem Klang das Hammergeräusch beigemischt …; trotzdem bezeichnet man den Klavierschall im Sprachgebrauch stets als Klang." In the *Handbuch der Physik*, Trendelenburg argued in the lines of Helmholtz that a treatment of acoustics separate from general mechanics was justified by its importance for general culture, and in more recent times, technical issues as well (Trendelenburg 1927, 1).

27. Eccles (1929). For Eccles, see Ratcliffe (1971). In 1918, Eccles and Frank Wilfred Jordan had patented a trigger circuit based on vacuum tubes, which later became known as the first flip-flop circuit.

28. Trendelenburg (1934, 77); see F. V. Hunt (1954, 66) for Wegel. Crandall introduced a range of electrical analogies in his *Theory of Vibrating Systems and Sound* (Crandall 1926).

29. J. Q. Stewart (1922). The astrophysicist John Stewart is better known for his later engagement with social physics. During World War I, he served as chief instructor in sound ranging at the Army Engineering School, after which he worked for AT&T until 1921 (see DeVorkin 2000, 208). This was before Bell Laboratories was created as a separate research laboratory for Western Electric and AT&T in 1925.

30. See Ferguson (1992, 137–147) for Polhem's mechanical alphabet and Reuleaux's machine grammar.

References

Archives

Archive of the Heinrch-Hertz-Institut für Schwingsforschung, Heinrich-Hertz-Institut für Nachrichtentechnik, Berlin

Bundesarchiv-Militärarchiv (BArch), Freiburg im Breisgau, Germany

Deutsches Museum Bildarchiv, Munich

Niels Bohr Library, Center for History of Physics, American Institute of Physics, College Park, MD

Science Museum Archive (SciMus), Science Museum, London

Nachlass Karl Willy Wagner, Archiv der Berlin-Brandenburgischen Akademie der Wissenschaften, Berlin

Nachlass Zenneck, Deutsches Museum Archiv, Munich

Siemens Corporate Archives, Munich/Berlin

SUB Göttingen Math. Archiv (Niedersächsische Staats- und Universitätsbibiothek Göttingen Mathematisches Archiv)

Telefunken Archiv, Historisches Archiv, Deutsches Technikmuseum Berlin

Universitätsarchiv Göttingen

Universitätsarchiv Stuttgart

Museum Collections

Hydrophones and Cambridge Scientific sound ranging apparatus, collected by Ernest Lancaster Jones in 1921 and 1922, Science Museum, London

Sound amplification system of the Deutsches Museum of 1925, registered in 1927, Deutsches Museum, Munich

Technisch-historische Sammlung "elektron" der Fakultät für Elektrotechnik, Technische Universität Dresden

Literature

Abraham, Otto, and Erich Moritz von Hornbostel. 1904. "Über die Bedeutung des Phonographen für die vergleichende Musikwissenschaft." *Zeitschrift fur Ethnologie* 36:222–236.

Adler, Stephen B., and Orville R. Butler. 1999. *Manufacturing the Future: A History of Western Electric.* Cambridge: Cambridge University Press.

Aigner, Franz. 1922. *Unterwasserschalltechnik. Grundlagen, Ziele und Grenzen— Submarine Akustik in Theorie und Praxis.* Berlin: Krayn.

Ames, Eric. 2003. "The Sound of Evolution." *Modernism/Modernity* 10 (2): 297–325.

Anduaga, Aitor. 2009. *Wireless and Empire: Geopolitics, Radio Industry, and Ionosphere in the British Empire, 1918–1939.* Oxford: Oxford University Press.

Angerer, Ernst von. "Ein registrierendes Saitengalvanometer von großer Registriergeschwindigkeit." *Zeitschrift für Instrumentenkunde* 17 (1): 1–6.

Angerer, Ernst von, and Rudolf Ladenburg. 1921. "Experimentelle Beiträge zur Ausbreitung des Schalles in der freien Atmosphäre." *Annalen der Physik* 371 (21): 293–322.

Ayrton, Hertha Marks. 1902. *The Electric Arc.* London: The Electrician.

Backhaus, Hermann. 1951. "Ernst Lübcke 60 Jahre." *Physikalische Blätter* 7 (1): 33.

Backhaus, Hermann, and Ferdinand Trendelenburg. 1925. "Akustische und physiologische Beobachtungen am Lautsprecher." *Wissenschaftliche Veröffentlichungen aus dem Siemens-Konzern* 4 (2): 205–208.

Barkan, Diana K. 1999. *Walther Nernst and the Transition to Modern Physical Science.* Cambridge: Cambridge University Press.

Barkhausen, Heinrich. 1907. *Das Problem der Schwingungserzeugung mit besonderer Berücksichtigung schneller elektrischer Schwingungen.* Leipzig: Hirzel.

Barkhausen, Heinrich. 1911. "Die Probleme der Schwachstromtechnik." *Dinglers Polytechnisches Journal* 326 (33): 513–517; (34): 531–534.

Barkhausen, Heinrich. 1923. *Elektronen-Röhren.* Leipzig: Hirzel.

Barkhausen, Heinrich. 1926. "Ein neuer Schallmesser für die Praxis." *Zeitschrift für technische Physik* 7:599–601.

Barkhausen, Heinrich, and G. Lewicki. 1924. "Die Empfindlichkeit des Ohres für nicht sinusförmige Töne." *Physikalische Zeitschrift* 25:537–541.

Barkhausen, Heinrich, and Hugo Lichte. 1920. "Quantitative Unterwasserschallversuche." *Annalen der Physik* 367 (14): 485–516.

Barkhausen, Heinrich, and Horst Tischner. 1927. "Die Lautstärke von zusammengesetzten Tönen und Geräuschen." *Zeitschrift für technische Physik* 8:215–221.

Barrow, Wilmer L. 1931. "Untersuchungen über den Heulsummer." *Annalen der Physik* 403 (2): 147–176.

Bartlett, Frederic Charles. 1927. *Psychology and the Soldier.* Cambridge: Cambridge University Press.

Bartlett, Frederic Charles. 1934. *The Problem of Noise.* Cambridge: Cambridge University Press.

Bausch, Wilhelm. 1939. *Schalldämmungsmessungen im Laboratorium und in fertigen Gebäuden.* Beihefte zum Gesundheits-Ingenieur 2, vol. 20. Munich: Oldenbourg.

Beetz, Wilhelm von. 1868. "Elektrisches Vibrations-Chronoskop." *Annalen der Physik* 211 (9): 126–134.

Benjamin, Walter. (1936) 1963. *Das Kunstwerk im Zeitalter seiner technischen Reproduzierbarkeit.* Frankfurt am Main: Suhrkamp.

Berg, Reno, and Johan Holtsmark. 1935a. "Akustiske målinger i en del forsamlingslokaler i Norge." In *Avhandlinger til 25 års jubileet,* 611–632. Trondheim: Norges Tekniske Høiskole. Also published in *Det Kongelige Norske Videnskabers Selskabs Skrifter,* no. 32 (1935).

Berg, Reno, and Johan Holtsmark. 1935b "Die Schallisolation von Doppelwänden I. Holzwände." *Det Kongelige Norske Videnskabers Selskabs Forhandlinger* 8 (23): 75–78.

Berger, Richard. 1911. *Über die Schalldurchlässigkeit.* Munich: Oldenbourg.

Berger, Richard. 1918. *Vorträge zur Einführung in das Schallmeßverfahren und die Arbeit des Schallmeßtrupps. Nach den an Stabsoffiziere der Artillerie gehaltenen Vorträgen.* Wahn: Artillerie-Messschule, July. (A copy can be found in the archive of the Deutsches Museum, Nachlass Zenneck, NL 053, folder 0293.)

Berger, Richard. 1926. *Die Schalltechnik.* Braunschweig: Vieweg and Sohn.

Berger, Richard. 1934. "Die Abwehr von Lärm und Erschütterungen." In Erich Waetzmann, ed., *Handbuch der Experimentalphysik: Technische Akustik,* vol. 2, 1–81. Leipzig: Akademische Verlagsgesellschaft.

Bergius, Rudolf. 1979. "Köhler, Wolfgang." *Neue Deutsche Biographie* 12:302–304.

Beyer, Robert T. 1999. *Sounds of Our Times—Two Hundred Years of Acoustics.* New York: Springer.

Beyerchen, Alan D. 1977. *Scientists under Hitler: Politics and the Physics Community in the Third Reich.* New Haven: Yale University Press.

Bijker, W. E., T. P. Hughes, and T. J. Pinch, eds. 1987. *The Social Construction of Technological Systems: New Directions in the Sociology and History of Technology*. Cambridge, MA: MIT Press.

Bijsterveld, Karin. 2008. *Mechanical Sound: Technology, Culture, and Public Problems of Noise in the Twentieth Century*. Cambridge, MA: MIT Press.

Birdsall, Carolyn. 2012. *Nazi Soundscapes: Sound, Technology and Urban Space, 1933–1945*. Amsterdam: Amsterdam University Press.

Bloor, David. 2000. "Whatever Happened to 'Social Constructiveness.'" In Akiko Saito, ed., *Bartlett, Culture and Cognition*, 194–215. Hove, UK: Psychology Press.

Bochow, Martin. 1933. *Schallmeßtrupp 51. Vom Krieg der Stoppuhren gegen Mörser und Haubitzen*. 2nd ed. Stuttgart: Union Deutsche Verlagsgesellschaft.

Boedeker, Karl, and Hans Riegger. 1920. "Über Bau und Anwendung eines mechanischen Schwingers." *Wissenschaftliche Veröffentlichungen aus dem Siemens-Konzern* 1 (1): 141–142.

Boersma, Kees. 2002. *Inventing Structures for Industrial Research: A History of the Philips Natlab 1914–1946*. Amsterdam: Aksant.

Borgnis, Fritz. 1961. "Fischer, Fritz." *Neue Deutsche Biographie* 5:185–186.

Braun, Hans-Joachim. 1998. "Lärmbelastung und Lärmbekämpfung in der Zwischenkriegszeit." In Günter Bayerl and Wolfhard Weber, eds., *Sozialgeschichte der Technik. Ulrich Troitzsch zum 60. Geburtstag*, 251–258. Münster: Waxmann.

Braun, Hans-Joachim. 2004. "Review Essay: Modern Sounds." *Social Studies of Science* 34 (5): 809–817.

Braunschweigische Wissenschaftliche Gesellschaft. 1969. "Mitglieder der BWG–Tischner, Horst." *Abhandlungen der Braunschweigischen Wissenschaftlichen Gesellschaft* 21:201–202.

Brenni, Paolo. 2011. "The Evolution of Teaching Instruments and Their Use between 1800 and 1930." In Peter Heering and Roland Wittje, eds., *Learning by Doing: Experiments and Instruments in the History of Science Teaching*, 281–315. Stuttgart: Franz Steiner.

Brüche, Ernst. 1938. *Zehn Jahre Forschungsinstitut der AEG*. Berlin-Reinickendorf.

Buchwald, J. Z., ed. 1995. *Scientific Practice—Theories and Stories of Doing Physics*. Chicago, London: University of Chicago Press.

Buchwald, Jed Z. 2013. "Electrodynamics from Thomson to Maxwell and Hertz." In Jed Z. Buchwald and Robert Fox, eds., *The Oxford Handbook of the History of Physics*, 571–583. Oxford: Oxford University Press.

Bürck, Werner, Paul Kotowski, and Hugo Lichte. 1935a. "Der Aufbau des Tonhöhen-bewußtseins." *Elektrische Nachrichten-Technik* 12 (10): 326–333.

Bürck, Werner, Paul Kotowski, and Hugo Lichte. 1935b. "Die Hörbarkeit von Laufzeitdifferenzen." *Elektrische Nachrichten-Technik* 12 (11): 355–362.

Busse, Detlef. 2008. *Engagement oder Rückzug? Göttinger Naturwissenschaften im Ersten Weltkrieg.* Göttingen: Universitätsverlag.

Cahan, David. 1989. *An Institute for an Empire: The Physikalisch-Technische Reichsanstalt, 1871–1918.* Cambridge: Cambridge University Press.

Carson, Cathryn, Alexei Kojevnikov, and Helmuth Trischler, eds. 2011. *Weimar Culture and Quantum Mechanics: Selected Papers by Paul Forman and Contemporary Perspectives on the Forman Thesis.* Singapore: World Scientific.

Chessa, Luciano. 2012. *Luigi Russolo, Futurist: Noise, Visual Arts, and the Occult.* Berkeley: University of California Press.

Cornell, Thomas D. 1986. "Merle A. Tuve and His Program of Nuclear Studies at the Department of Terrestrial Magnetism: The Early Career of a Modern American Physicist." PhD diss., Johns Hopkins University.

Crafton, Donald. 1997. *The Talkies: American Cinema's Transition to Sound, 1926–1931.* New York: Charles Scribner's Sons.

Crandall, Irving B. 1926. *Theory of Vibrating Systems and Sound.* New York: D. Van Nostrand.

Cremer, Lothar. 1972. "Erwin Meyer 21. Juli 1899–6. März 1972." *Jahrbuch der Akademie der Wissenschaften in Göttingen* 1972:179–185.

Crone, Willy, Hans Seiberth, and Jonathan Zenneck. 1934. "Die Verbesserung der Akustik im Prinzregententheater München." *Annalen der Physik* 411 (3): 299–304.

Darrigol, Olivier. 2000. *Electrodynamics from Ampère to Einstein.* Oxford: Oxford University Press.

Daston, Lorraine, and Peter Galison. 1992. "The Image of Objectivity." *Representations (Berkeley, Calif.)* 40: 81–128.

Dennis, Michael Aaron. 1991, "A Change of State: The Political Cultures of Technical Practice at the MIT Instrumentation Laboratory and the Johns Hopkins University Applied Physics Laboratory, 1930–1945." PhD diss., Johns Hopkins University.

DEPATISnet. 2015. (Deutsches Patent- und Markenamt patent database; accessed 27 May). https://depatisnet.dpma.de/DepatisNet/.

Des Coudres, Theodor. 1919. "Hermann Th. Simon †." *Physikalische Zeitschrift* 20 (14): 313–320.

Deutsche Akademie der Luftfahrtforschung. 1939. *Schriften der Deutschen Akademie der Luftfahrtforschung*. Heft 10. Vorträge gehalten in der 7. Wissenschaftssitzung der ordentlichen Mitglieder am 10. Februar 1939. Berlin: Kommissionsverlag von R. Oldenbourg.

DeVorkin, David H. 2000. *Henry Norris Russell: Dean of American Astronomers*. Princeton, N.J.: Princeton University Press.

"Die größte Lautsprecheranlage der Welt." 1933. *Funk Bastler* 20: 308.

"Die Telefunken-Großlautsprecheranlage auf dem Tempelhofer Feld." 1933. *Telefunken-Zeitung* 63: 53–55.

Distelmeyer, J., ed. 2003. *Tonfilmfrieden / Tonfilmkrieg: Die Geschichte der Tobis vom Technik-Syndikat zum Staatskonzern*. Munich: Edition Text und Kritik.

Doetsch, Carl W. H. 1920. "Kamina und das Los der Togogefangenen." *Telefunken-Zeitung* 19: 29–41.

Donhauser, Peter. 2007. *Elektrische Klangmaschinen—Die Pionierzeit in Deutschland und Österreich*. Wien: Böhlau.

Dörfel, Günter, and Dieter Hoffmann. 2005. *Von Albert Einstein bis Norbert Wiener—frühe Ansichten und späte Einsichten zum Phänomen des elektronischen Rauschens*. Berlin: Max Planck Institute for the History of Science Preprint 301.

"Dr. Heinrich Hecht zum 60. Geburtstag." 1940. *Akustische Zeitschrift* 5 (1): 1–2.

Duddell, William Du Bois. 1900. "On Rapid Variations in the Current through the Direct-Current Arc." *Journal of the Institution of Electrical Engineers* 30:232–267. Also published that same year in *Electrician* 46:269–273, 310–313.

Eccles, William Henry. 1929. "The New Acoustics." *Proceedings of the Physical Society* 41:231–239.

Eckert, Michael. 2006. *The Dawn of Fluid Dynamics*. Weilheim: Wiley.

Eckert, Michael, and Karl Märker, eds. 2000. *Arnold Sommerfeld—Wissenschaftlicher Briefwechsel*. vol. 1, 1892–1918. Berlin and Munich: GNT-Verlag and Deutsches Museum.

Eddington, Arthur S. 1923. *The Mathematical Theory of Relativity*. Cambridge: Cambridge University Press.

Edgerton, David. 2004. "'The Linear Model' Did Not Exist: Reflections on the History and Historiography of Science and Research in Industry in the Twentieth Century." In Karl Grandin and Nina Wormbs, ed., *The Science–Industry Nexus: History, Policy, Implications*, 31–57. Sagamore Beach, MA: Science History Publications.

Edgerton, David. 2007. *The Shock of the Old: Technology and Global History since 1900*. Oxford: Oxford University Press.

Eggenschwiler, Jurt, and Sabine von Fischer. 2014. "Geschichte der Akustik an der ETH Zürich und an der Empa Dübendorf." Fortschritte der Akustik—DAGA 2014 Oldenburg: 764–765.

Ellis, Alexander J. 1885. "On the Musical Scales of Various Nations." *Journal of the Society of Arts* 33:485–527.

Elshakry, Marwa. 2010. "When Science Became Western: Historiographical Reflections." *Isis* 101 (1): 98–109.

Emde, Hertha, Hans E. Heinrich, and Oskar Vierling. 1937. "Ein Beitrag zum Problem der Großschallübertragungsanlagen." *Zeitschrift für technische Physik* 18 (9): 252–255.

Encke, Julia. 2006. *Augenblicke der Gefahr: Der Krieg und die Sinne 1914–1934.* Paderborn: Wilhelm Fink Verlag.

Engel, Friedrich K. 1999. The Introduction of the Magnetophon. In Eric D. Daniel, C. Denis Mee, and Mark H. Clark, eds., *Magnetic Recording: The First 100 Years*, 47–71. New York: IEEE.

Engel, Michael. 1997. "Narath, Albert." *Neue Deutsche Biographie* 18:734–735.

Engl, Jo. (Josef). 1927. *Der Tönende Film—Das Triergon-Verfahren und seine Anwendungsmöglichkeiten.* Braunschweig: Vieweg and Sohn.

Erker, Paul. 1995. "The Choice between Competition and Cooperation: Research and Development in the Electrical Industry in Germany and the Netherlands, 1926–1936." In François Caron, Paul Erker, and Wolfram Fischer, eds., *Innovations in the European Economy between the Wars*, 231–253. Berlin: Walter de Gruyter.

Esau, Abraham. 1919. "Die Großstation Kamina und der Beginn des Weltkrieges." *Telefunken-Zeitung* 16:31–36.

Eyman, Scott. 1997. *The Speed of Sound: Hollywood and the Talkie Revolution, 1926–1930.* New York: Simon and Schuster.

Fassbender, Heinrich, and Kurt Krüger. 1927. "Geräuschmessung in Flugzeugen." *Zeitschrift für technische Physik* 8:277–282.

Faxén, Hilding, and Johan Holtsmark. 1927. "Beitrag zur Theorie des Durchganges langsamer Elektronen durch Gase." *Zeitschrift für Physik* 45 (5/6): 307–324.

Feldenkirchen, Wilfried. 1997. *Siemens. Von der Werkstatt zum Weltunternehmen.* Munich: Piper.

Felgel-Farnholz, Richard. 1983. "Schubert, Georg Oskar. 1900–1955." In Sigfrid von Weiher, ed., *Männer der Funktechnik*, 165–167. Offenbach: VDE-Verlag.

Ferguson, Eugene S. 1992. *Engineering and the Mind's Eye.* Cambridge, MA: MIT Press.

Fischer, Claude S. 1992. *America Calling: A Social History of the Telephone to 1940*. Berkeley: University of California Press.

Fischer, Fritz, and Hugo Lichte. 1931. *Tonfilm Aufnahme und Wiedergabe nach dem Klangfilm-Verfahren*. Leipzig: Hirzel.

Fischer, Fritz, and Hugo Lichte. 1934. "Der Tonfilm." In Erich Waetzmann, ed., *Handbuch der Experimentalphysik: Technische Akustik*, vol. 2, 349–408. Leipzig: Akademische Verlagsgesellschaft.

Fleming, John Ambrose. 1906. *The Principles of Electric Wave Telegraphy*. London: Longmans.

Fletcher, Harvey. 1923. "Physical Measurements of Audition and Their Bearing on the Theory of Hearing." *Journal of the Franklin Institute* 193:289–326.

Fletcher, Harvey. 1929. *Speech and Hearing*. New York: D. Van Nostrand.

Fletcher, Harvey. 1958. "George W. Stewart, 1876–1956." *Biographical Memoirs. National Academy of Sciences (U. S.)* 32:378–398.

Forbes, George. 1878. "The Telephone, an Instrument of Precision." *Nature* 17:343.

Forman, Paul. 1971. "Weimar Culture, Causality and Quantum Theory: Adoption by German Physicists and Mathematicians to a Hostile Intellectual Environment." *Historical Studies in the Physical Sciences* 3:1–115.

Forman, Paul. 1973. "Scientific Internationalism and the Weimar Physicists: The Ideology and Its Manipulation in Germany after World War I." *Isis* 64 (2): 150–180.

Freystedt, Erich. 1935. "Das 'Tonfrequenz-Spektrometer,' ein Frequenzanalysator mit äußerst hoher Analysiergeschwindigkeit und unmittelbar sichtbarem Spektrum." *Zeitschrift für technische Physik* 12:533–539.

Friedewald, Michael. 2002. "Funkentelegrafie und deutsche Kolonien: Technik als Mittel imperialistischer Politik." In Kai Handel, ed., *Von der Telegraphie zum Internet—Kommunikation in Geschichte und Gegenwart*, 51–63. Freiberg: Georg Agricola Gesellschaft.

Fritz, Walter. 1964. "Gehlhoff, Georg Richard." *Neue Deutsche Biographie* 6:135–136.

Fuchs, Franz. 1933. "Die Jubiläums-Funkausstellung in Berlin." *Hochfrequenztechnik und Elektroakustik* 42 (6): 208–212.

Fuchs, Franz. 1963. *Der Aufbau der technischen Akustik im Deutschen Museum*. Munich: R. Oldenbourg; Düsseldorf: VDI-Verlag.

Gabor, Dennis. 1946. "Theory of Communication." *Journal of the Institute of Electrical Engineers* 93, pt. 3 (26): 429–457.

Gabor, Dennis. 1947. "Acoustical Quanta and the Theory of Hearing." *Nature* 159:591–594.

Gehlhoff, Georg. 1921. "Die Ausbildung der technischen Physiker." *Zeitschrift für technische Physik* 2 (3): 121–127.

Gehlhoff, Georg, Hans Rukop and Wilhelm Hort. 1920. "Zur Einführung / Zur Gründung der Deutschen Gesellschaft für technische Physik." *Zeitschrift für technische Physik* 1:1–6.

Gemelli, Agostino, and Giuseppina Pastori. 1934. *L'analisi elettroacustica del linguaggio.* 2 vols. Milano: Vita e Pensiero.

Gerber, Stefan. 2009. "Die Universität Jena 1850–1918." In Senatskommission zur Aufarbeitung der Jenaer Universitätsgeschichte im 20. Jahrhundert, *Traditionen—Brüche—Wandlungen: Die Universität Jena 1850–1995*, 23–269. Cologne: Böhlau.

Gerdien, Hans. 1922. "Über einen akustischen Schwinger (Nach gemeinsamen mit H. Riegger und K. Boedeker ausgeführten Versuchen)." *Zeitschrift für technische Physik* 3 (2): 40–44.

Gerdien, Hans. 1926. "Hans Riegger †." *Zeitschrift für technische Physik* 7:321–324.

Gerdien, Hans, and Hans Riegger. 1920. "Ein akustischer Schwinger." *Wissenschaftliche Veröffentlichungen aus dem Siemens-Konzern* 1 (1): 137–140.

Gibling, Sophie P. 1917. "Types of Musical Listening." *Musical Quarterly* 3 (3): 385–389.

Gieryn, Thomas F. 1983. "Boundary-Work and the Demarcation of Science from Non-science: Strains and Interests in Professional Ideologies of Scientists." *American Sociological Review* 48:781–795.

Gieryn, Thomas F. 1999. *Cultural Boundaries of Science: Credibility on the Line.* Chicago: University of Chicago Press.

Goetzeler, Herbert. 1994. "Gustav Hertz 1887–1975." In Ernst Feldtkeller and Herbert Goetzeler, eds., *Pioniere der Wissenschaft bei Siemens*, 78–84. Erlangen: Publicis MCD Verlag.

"Gründung einer 'Gesellschaft für elektrische Musik.'" 1932. *Funk* 44:176.

Grützmacher, Martin, and Erwin Meyer. 1936. "Zur Einführung." *Akustische Zeitschrift* 1:1.

Guicking, Dieter. 2012. *Erwin Meyer—Ein bedeutender deutscher Akustiker: Biographische Notizen.* Göttingen: Universitätsverlag.

Gundlach, Friedrich-Wilhelm. 1978. "Das Heinrich-Hertz-Institut für Schwingungsforschung." In *50 Jahre Heinrich-Hertz-Institut—Vortragsband*, 4–16. Berlin: Heinrich-Hertz-Institut für Nachrichtentechnik.

Günther, Siegmund. 1902. "Akustisch-geographische Probleme." In *Sitzungsberichte der mathematisch-physikalischen Classe der königlich bayerischen Akademie der Wissenschaften*, vol. 31, meetings of January 5 and July 6, 1901, 15–33 and 211–263. Munich: Verl. d. K. Akad.

Haber, Fritz. 1924. *Fünf Vorträge aus den Jahren 1920–1923*. Berlin: Springer.

Hackmann, Willem Dirk. 1984. *Seek and Strike: Sonar, Anti-submarine Warfare, and the Royal Navy, 1914–54*. London: HMSO.

Hackmann, Willem Dirk. 1986. "Sonar Research and Naval Warfare, 1914–1954: A Case Study of a Twentieth-Century Science." *Historical Studies in the Physical and Biological Sciences* 16 (1): 83–110.

Hagemeyer, Friedrich Wilhelm. 1979. "Die Entstehung von Informationskonzepten in der Nachrichtentechnik—Eine Fallstudie zur Theoriebildung in der Technik in Industrie- und Kriegsforschung."PhD diss., Freie Universität Berlin.

Hahnemann, Walter. 1922. "Schwingungstechnische Probleme als Grundlage der technischen Akustik." *Zeitschrift für technische Physik* 3 (2): 44–46.

Hahnemann, Walter, and Heinrich Hecht. 1916. "Schallfelder und Schallantennen." *Physikalische Zeitschrift* 17 (24): 601–609.

Hahnemann, Walter, and Heinrich Hecht. 1919a. "Schallgeber und Schallempfänger I." *Physikalische Zeitschrift* 20:104–114.

Hahnemann, Walter, and Heinrich Hecht. 1919b. "Schallgeber und Schallempfänger II." *Physikalische Zeitschrift* 20:245–251.

Hahnemann, Walter, and Heinrich Hecht. 1919c. "Der mechanisch-akustische Aufbau eines Telephons." *Annalen der Physik* 365:454–480.

Hahnemann, Walter, and Hugo Lichte. 1920. "Die moderne Entwicklung der Unterwasserschalltechnik in Deutschland." *Naturwissenschaften* 8 (45): 871–878.

Handbook of the Sound Ranging Instrument. 1921. London: Harrison and Sons for HMSO.

Harbeck, R. 1943a. *Beiträge zur Geschichte des Schall- und Lichtmessens*. Idar-Oberstein: Gesellschaft für Artilleriekunde Idar-Oberstein.

Harbeck, R. 1943b. Ergänzungs- und Änderungsblatt 1. Typoskript. *Beiträge zur Geschichte des Schall- und Lichtmessens*. Idar-Oberstein: Gesellschaft für Artilleriekunde Idar-Oberstein.

Hars, Florian. 1999. *Ferdinand Braun 1850–1918. Ein Wilhelminischer Physiker*. Berlin: GNT-Verlag.

Hartcup, Guy. 1988. *The War of Invention: Science in the Great War, 1914–18*. London: Brassey's Defence Publishers.

Hartmann and Braun. 1886. "Nippoldt's Telephonbrücke zum Messen von Erdlei-tungswiderständen." *Dinglers Polytechnisches Journal* 261:202–204.

Hartmann and Braun. 1894. *Instruments de mesure électriques*. Frankfurt am Main: Hartmann and Braun.

Hartmann, Carl A. 1922. "Über die Bestimmung des elektrischen Elementarquan-tums aus dem Schroteffekt." *Annalen der Physik* 362:51–78.

Hartmann-Kempf, Robert. 1903. *Über den Einfluß der Amplitude auf Tonhöhe und Decrement von Stimmgabeln und zungenförmigen Stahlfedern. Elektroakustische Untersuc-hungen*. Frankfurt am Main: Knauer.

Hartmann-Kempf, Robert. 1904. "Über den Einfluß der Amplitude auf die Tonhöhe und das Dekrement von Stimmgabeln und zungenförmigen Stahlfederbändern." *Annalen der Physik* 318:124–162.

Hashagen, Ulf. 2003. *Walther von Dyck (1856–1934): Mathematik, Technik und Wissenschaftsorganisation an der TH München*. Stuttgart: Franz Steiner.

Hecht, Heinrich. 1939. *Schaltschemata und Differentialgleichungen elektrischer und mechanischer Schwingungsgebilde*. Leipzig: Johann Ambrosius Barth.

Hecht, Heinrich. 1941. *Die elektroakustischen Wandler*. Leipzig: Johann Ambrosius Barth.

Hecht, Heinrich. 1954. *Vier Fragen an den Weltäther*. Göttingen: Musterschmidt.

Hecht, Heinrich, and F. A. Fischer. 1934. "Anwendung der Ausbreitung des Schalles in freien Medien." In Erich Waetzmann, ed., *Handbuch der Experimentalphysik: Technische Akustik*, vol. 1, 355–442. Leipzig: Akademische Verlagsgesellschaft.

Heger, Richard. 1911. "Zur Theorie und Praxis der Raumakustik." *Zeitschrift für Architektur und Ingenieurwesen* 16/57 (4): 309–322.

Hehlgans, Friedrich Wilhelm. 1928. "Über Piezoquarzplatten als Sender und Empfänger hochfrequenter akustischer Schwingungen." *Annalen der Physik* 391 (12): 587–627.

Heidelberger, Michael. 1994. "Helmholtz' Erkenntnis- und Wissenschaftstheorie im Kontext der Philosophie und Naturwissenschaft des 19. Jahrhunderts." In Korenz Krüger, ed., *Universalgienie Helmholtz: Rückblick nach 100 Jahren*, 168–185. Berlin: Akademie Verlag.

Heilbron, John L., and Robert W. Seidel. 1989. *Lawrence and His Laboratory: A History of the Lawrence Berkeley Laboratory*. Berkeley: University of California Press.

Heinrich-Hertz-Institut für Schwingungsforschung. 1933. *Bericht über die Tätigkeit des Instituts im Geschäftsjahr 1932/33*. Berlin: Heinrich-Hertz-Institut.

Helmholtz, Hermann von. 1853. "Ueber einige Gesetze der Vertheilung elektrischer Ströme in körperlichen Leitern mit Anwendung auf die thierisch-elektrischen Versuche." *Annalen der Physik und Chemie* 165 (6): 211–233.

Helmholtz, Hermann von. 1856. "Ueber Combinationstöne." *Annalen der Physik und Chemie* 175 (12): 497–540.

Helmholtz, Hermann von. 1863. *Die Lehre von den Tonempfindungen als physiologische Grundlage für die Theorie der Musik.* Braunschweig: Vieweg and Sohn.

Helmholtz, Hermann von. 1867. *Handbuch der physiologischen Optik.* Leipzig: Leopold Voss.

Helmholtz, Hermann von. 1870. *Die Lehre von den Tonempfindungen als physiologische Grundlage für die Theorie der Musik.* 3rd ed. Braunschweig: Vieweg and Sohn.

Helmholtz, Hermann von. (1863) 1875. *On the Sensations of Tone as a Physiological Basis for the Theory of Music.* Translated from the 3rd German edition by Alexander J. Ellis. London: Longmans.

Helmholtz, Hermann von. (1862) 1896. "Ueber das Verhältniss der Naturwissenschaften zur Gesamtheit der Wissenschaft: Akademische Festrede gehalten zu Heidelberg beim Antritt des Prorectorats 1862." In Hermann von Helmholtz, *Vorträge und Reden, vierte Auflage,* vol. 1, 157–185. Braunschweig: Vieweg and Sohn.

Helmholtz, Hermann von. (1867) 1925. *Helmholtz's Treatise of Physiological Optics,* vol. 3, *The Perceptions of Vision.* Translated from the 3rd German edition. Edited by James P. C. Southall. Menasha, WI: Optical Society of America.

Hentschel, Klaus. 2007. *The Mental Aftermath: The Mentality of German Physicists 1945–1949.* Oxford: Oxford University Press.

Hentschel, K., ed. 2010. *Analogien in Naturwissenschaften, Medizin und Technik.* Stuttgart: Wissenschaftliche Verlagsges.

Hertz, Heinrich. 1893. *Electric Waves: Researches on the Propagation of Electric Action with Finite Velocity through Space.* London: MacMillan.

Hiebert, Erwin. 2014. *The Helmholtz Legacy in Physiological Acoustics.* Cham: Springer.

Hoffmann, Christoph. 1994. "Wissenschaft und Militär: Das Berliner psychologische Institut und der 1. Weltkrieg." *Psychologie und Geschichte* 5 (3/4): 261–285.

Hoffmann, Christoph. 2003. "Helmholtz' Apparatuses: Telegraphy as Working Model of Nerve Physiology." *Philosophia Scientiae* 7 (1): 129–149.

Hoffmann, Dieter. 1993. "Nationalsozialistische Gleichschaltung und Tendenzen militärischer Forschungsorientierung an der Physikalisch-Technischen Reichsanstalt im Dritten Reich." In Helmut Albrecht, ed., *Naturwissenschaft und Technik in der Geschichte,* 121–131. Stuttgart: GNT-Verlag.

Hoffmann, Dieter, and Edgar Swinne. 1994. *Über die Geschichte der "technischen Physik" in Deutschland und den Begründer ihrer wissenschaftlichen Gesellschaft Georg Gehlhoff*. Berlin: ERS-Verl.

Hong, Sungook. 2001. *Wireless: From Marconi's Black-Box to the Audion*. Cambridge, MA: MIT Press.

Honigsheim, Paul. 1930. "Musik und Gesellschaft." In Leo Kestenberg, ed., *Kunst und Technik*, 63–96. Berlin: Volksverband der Bücherfreunde.

Hormann, Ernst. 1932. "Zur Theorie der magnetischen Tonaufzeichnung." *Elektrische Nachrichten-Technik* 9 (10): 388–403.

Hornbostel, Erich Moritz von. 1905. "Die Probleme der vergleichenden Musikwissenschaft." *Zeitschrift der internationalen Musikgesellschaft* 7 (3): 85–97.

Hughes, Ivor. 2009. "Professor David Edward Hughes." *AWA Review* 22:111–134.

Hughes, Jeff. 1993. "The Radioactivists: Community, Controversy and the Rise of Nuclear Physics." PhD diss., University of Cambridge.

Hughes, Jeff. 1998. "Plasticine and Valves: Industry, Instrumentation and the Emergence of Nuclear Physics." In Jean-Paul Gaudillière and Ilana Löwy, eds., *The Invisible Industrialist—Manufacturers and the Production of Scientific Knowledge*, 58–101. Basingstoke, NY: Macmillan and St. Martin's.

Hughes, Thomas P. 1983. *Networks of Power: Electrification in Western Society, 1880–1930*. Baltimore: Johns Hopkins University Press.

Hughes, Thomas P. 1987. "The Evolution of Large Technological Systems." In Wiebe E. Bijker, Thomas P. Hughes, and Trevor J. Pinch, eds., *The Social Construction of Technological Systems: New Directions in the Sociology and History of Technology*, 51–82. Cambridge, MA: MIT Press.

Hui, Alexandra. 2013. *The Psychophysical Ear: Musical Experiments, Experimental Sounds, 1840–1910*. Cambridge, MA: MIT Press.

Hunke, Heinrich. (1933) 1935. *Luftgefahr und Luftschutz. Mit besonderer Berücksichtigung des deutschen Luftschutzes*. 2nd ed. Berlin: E. S. Mittler and Sohn.

Hunt, Bruce. 1991. *The Maxwellians*. Ithaca, NY: Cornell University Press.

Hunt, Frederick Vinton. 1954. *Electroacoustics: The Analysis of Transduction, and Its Historical Background*. Cambridge, MA.: Harvard University Press.

Institut für Schwingungsforschung der Technischen Hochschule Berlin. 1938. *Tätigkeitsbericht über das Berichtsjahr 1938*. Berlin: Institut für Schwingungsforschung.

Institut für Schwingungsforschung der Technischen Hochschule Berlin. 1940. *Tätigkeitsbericht über die Berichtsjahr2 1939/40*. Berlin: Institut für Schwingungsforschung.

Jack, Evan Maclean. 1920. *Report on Survey on the Western Front 1914–1918*. London: HMSO.

Jackson, Myles W. 2006. *Harmonious Triads—Physicists, Musicians, and Instrument Makers in Nineteenth-Century Germany*. Cambridge, MA: MIT Press.

Jagwitz, Fritz von. 1918. "Aufnahme von Schallschwingungen mittels des Oszillographen." *Mitteilungen der Flak-Scheinwerfer-Prüf-und Versuchsabteilung Hannover* 3:3–13.

Jahrbuch des Forschungs-Instituts der Allgemeinen Elektricitäts-Gesellschaft. 1930–1938. Vol. 1 (1928–29)–vol. 5 (1936–37). Berlin: Springer.

Joerges, B., and T. Shinn, eds. 2000. *Instrumentation: Between Science, State and Industry*. Dordrecht: Kluwer Academics.

Johnson, Don H. 2003a. "Origins of the Equivalent Circuit Concept: The Voltage-Source Equivalent." *Proceedings of the IEEE* 91:636–640.

Johnson, Don H. 2003b. "Origins of the Equivalent Circuit Concept: The Current-Source Equivalent." *Proceedings of the IEEE* 91:817–821.

Johnson, Jon Bertrand. 1925. "The Schottky Effect in Low Frequency Circuits." *Physical Review* 26:71–85.

Jones-Imhotep, Edward. 2008. "Icons and Electronics." *Historical Studies in the Natural Sciences* 38 (3): 405–450.

Joos, Georg. 1953. "Angerer, Lorenz Ludwig Maximilian Ernst von." *Neue Deutsche Biographie* 1: 292.

Jossé, Harald. 1984. *Die Entstehung des Tonfilms—Beitrag zu einer faktenorientierten Mediengeschichtsschreibung*. Freiburg: Verlag Karl Alber.

Jung, Gerhard. 1961. "Fredenhagen, Karl Hermann." *Neue Deutsche Biographie* 5: 386.

Kaiser, Walter. 1993. "Helmholtz's Instrumental Role in the Foundation of Classical Electrodynamics." In David Cahan, ed., *Hermann von Helmholtz and the Foundations of Nineteenth-Century Science*, 374–402. Berkeley: University of California Press.

Kapp, Gisbert. 1907. *Transformatoren für Wechselstrom und Drehstrom*. Berlin: Julius Springer.

Katzir, Shaul. 2006. *The Beginnings of Piezoelectricity: A Study in Mundane Physics*. Boston Studies in Philosophy of Science 246. Dordrecht: Springer.

Katzir, Shaul. 2010. "War and Peacetime Research on the Road to Crystal Frequency Control." *Technology and Culture* 51 (1): 99–125.

Katzir, Shaul. 2012. "Who Knew Piezoelectricity? Rutherford and Langevin on Submarine Detection and the Invention of Sonar." *Notes and Records of the Royal Society* 66:141–157.

Kaufmann, Stefan. 1996. *Kommunikationstechnik und Kriegsführung—Stufen telemedialer Rüstung*. Munich: Wilhelm Fink Verlag.

Kempf-Hartmann, Robert. 1902. "Photographische Darstellung der Schwingungen von Telephonmembranen." *Annalen der Physik* 313 (7): 481–538.

Kern, Ulrich. 1994. *Forschung und Präzisionsmessung: Die Physikalisch-Technische Reichsanstalt zwischen 1918 und 1948*. Weinheim: VCH.

Kestenberg, L., ed. 1930. *Kunst und Technik*. Berlin: Volksverband der Bücherfreunde.

Kevles, Daniel J. 1978. *The Physicists: A History of a Scientific Community in Modern America*. New York: Knopf.

Kline, Ronald. 1992. *Steinmetz: Engineer and Socialist*. Baltimore: Johns Hopkins University Press.

Kluge, Martin. 1933. "Problem der Dämpfung des Auspuffschalles der Kraftfahrzeugmotoren." *Automobiltechnische Zeitschrift* 36 (7): 192–196; (9): 244–249.

Kluge, Martin. 1941. "H. Barkhausens Beiträge zur akustischen Forschung." *Akustische Zeitschrift* 6 (6): 313–318.

Knoblauch, Oscar. 1941–1942. *Die Geschichte des Laboratoriums für technische Physik der Technischen Hochschule München 1902–1934*. Munich: Technische Hochschule.

Koch, Hans Jürgen, and Hermann Glaser. 2005. *Ganz Ohr—eine Kulturgeschichte des Radios in Deutschland*. Cologne: Böhlau.

Koenig, Rudolph. 1876. "Ueber den Zusammenklang zweier Töne." *Annalen der Physik* 233:177–237.

Költzsch, P., ed. 2003. *Festschrift zum Ehrenkolloquium Reichardt—Kraak—Wöhle: Zur Entwicklung des Fachgebietes Technische Akustik und des Akustischen Instituts an der Technischen Hochschule/Technischen Universität Dresden in den Zeitläufen des 20. Jahrhunderts*. Dresden: Technische Universität.

König, Wolfgang. 1993. "Technical Education and Industrial Performance in Germany: A Triumph of Heterogeneity." In Robert Fox and Anna Guagnini, eds., *Education, Technology and Industrial Performance in Europe, 1850–1939*, 65–88. Cambridge: Cambridge University Press.

König, Wolfgang. 1995. *Technikwissenschaften—Die Entstehung der Elektrotechnik aus Industrie und Wissenschaft zwischen 1880 und 1914*. Chur: G+B Verlag Fakultas.

König, Wolfgang. 2003a. "Mythen um den Volksempfänger. Revisionistische Untersuchungen zur nationalsozialistischen Rundfunkpolitik." *Technikgeschichte* 70 (2): 73–102.

König, Wolfgang. 2003b. "Der Volksempfänger und die Radioindustrie. Ein Beitrag zum Verhältnis von Wirtschaft und Politik im Nationalsozialismus." *Vierteljahrschrift für Sozial- und Wirtschaftsgeschichte* 90 (3): 269–289.

König, Wolfgang. 2004. *Volkswagen, Volksempfänger, Volksgemeinschaft. "Volksprodukte" im Dritten Reich. Vom Scheitern einer nationalsozialistischen Konsumgesellschaft.* Paderborn: Ferdinand Schöningh.

Königsberger, Leo. 1902–1903. *Hermann von Helmholtz.* 3 vols. Braunschweig: Vieweg and Sohn.

Kragh, Helge. 1999. *Quantum Generations—A History of Physics in the Twentieth Century.* Princeton, NJ: Princeton University Press.

Krieck, Ernst. 1936. *Nationalpolitische Erziehung.* 20th ed. Leipzig: Armanen.

Krug, Erich. 1935. "Vom Lärm im Betrieb–Die Reichswochenschau ohne Lärm vom 6. bis 12. Mai 1935." *Stahl und Eisen* 55 (20): 548–549.

Krüger, Friedrich. 1921. "Die Stellung und das Studium der physikalisch-mathematischen Wissenschaften an den deutschen Hochschulen." *Zeitschrift für technische Physik* 2 (5): 113–121.

Küchenmeister, Heinrich J. 1930. "Tonfilm als Faktor in Technik, Wirtschaft und Kultur." In Leo Kestenberg, ed., *Kunst und Technik,* 361–363. Berlin: Volksverband der Bücherfreunde.

Kuntze, Walter. 1930. "Beiträge zur Raumakustik." *Annalen der Physik* 396 (8): 1058–1096.

Kurz, T., U. Parlitz, and U. Kaatze, eds. 2007. *Oscillations, Waves and Interactions: Sixty Years Drittes Physikalisches Institut. A Festschrift.* Göttingen: Universitätsverlag.

Kurzel-Runtscheiner, Erich. 1953. "Aigner, Franz Johann." *Neue Deutsche Biographie* 1: 118.

Kuttruff, Heinrich. 1969. "Erwin Meyer zum 70. Geburtstag." *Physikalische Blätter* 25 (7): 324.

Ladenburg, Rudolf, and Ernst von Angerer. 1918. *Über die Ausbreitung des Schalles in der freien Atmosphäre.* Berlin: Preußische Artillerie-Prüfungs-Kommission.

Lancaster Jones, Ernest. 1922 "Sound-Ranging." *Proceedings of the Musical Association* 48:77–89.

Landé, Alfred. 1930. *Vorlesungen über Wellenmechanik.* Leipzig: Akademische Verlagsgesellschaft.

Lange, Britta. 2011. "South Asian Soldiers and German Academics: Anthropological, Linguistic and Musicological Field Studies in Prison Camps." In Ravi Ahuja, Heike Liebau, and Franziska Roy, eds., 'When the war began we heard of several kings'—South Asian Prisoners in World War I Germany, 149–184. Delhi: Social Science Press.

Lange, Gertrud. 1910. "Beiträge zur Kenntnis der Lichtbogenhysteresis." Annalen der Physik 337 (8): 589–647.

Le Corbeiller, Philippe. 1934. Electro-acoustique. Paris: Étienne Chiron.

Lembrecht, Christina. 2007. "Wissenschaftsverlage im Feld der Physik. Profile und Professionsverschiebungen 1900–1933." Archiv für Geschichte des Buchwesens 61:111–200.

Lenoir, Timothy. 1994. "Helmholtz and the Materialities of Communication." Osiris 9:184–207.

Lichte, Hugo. 1913. "Über die Schallintensität des tönenden Lichtbogens." Annalen der Physik 347 (14): 843–870.

Lichte, Hugo. 1919. "Über den Einfluss horizontaler Temperaturschichtung des Seewassers auf die Reichweite von Unterwasserschallsignalen." Physikalische Zeitschrift 17:385–389.

Lichte, Hugo. 1940. "Dr. Heinrich Hecht zum 60. Geburtstag." Zeitschrift für technische Physik 21 (2): 25–27.

Lichte, Hugo. 1965. "Der Tonfilm." In Bruno Schweder, ed., Forschen und schaffen. Beiträge der AEG zur Entwicklung der Elektrotechnik bis zum Wiederaufbau nach dem zweiten Weltkrieg, vol. 3, 421–427. Berlin: AEG.

Lilienthal-Gesellschaft für Luftfahrtforschung. 1937. Geräusch und Geräuschdämpfung. Bericht über die Sondertagung in Gottingen am 7./8. Dezember 1937. Bericht 088/011. Berlin: Lilienthal-Gesellschaft für Luftfahrtforschung.

Linck, Wolfgang. 1930. "Beiträge zur Raumakustik." Annalen der Physik 396 (8): 1017–1057.

Lindsay, Robert Bruce. 1945. "Introduction." In Lord Rayleigh (John William Strutt), The Theory of Sound, vol. 1, reprint of the 1894 edition, v–xxxii. New York: Dover.

Livingstone, David N. 2003. Putting Science in Its Place: Geographies of Scientific Knowledge. Chicago: University of Chicago Press.

Lodge, Oliver. 1880. "On Intermittent Currents and the Theory of the Induction-Balance." Philosophical Magazine Series 5 9 (54): 123–146.

Lodge, Oliver. 1889. Modern Views of Electricity. London: MacMillan.

Lorenz, Detlef. 2004. Das AEG-Forschungsinstitut in Berlin-Reinickendorf. Daten, Fakten, Namen zu seiner Geschichte 1928–1989. Berlin: privately printed.

Löwenstein, Leo. 1928. "Die Erfindung der Schallmessung." *Die Schalltechnik* 1 (2): 21–24.

Lübcke, Ernst. 1923. "Über einen wasserdichten Apparat zur Wahrnehmung von Flugzeugschall." *Zeitschrift für technische Physik* 4 (3): 99–101.

Lübcke, Ernst. 1931. "Das akustische Laboratorium das Physikalischen Instituts der Technischen Hochschule zu Braunschweig." *Die Schalltechnik* 4 (5): 78–80.

Lübcke, Ernst. 1943. "Karl Willy Wagners Beiträge zur akustischen Forschung." *Akustische Zeitschrift* 8:78–80.

Lübcke, Ernst. 1956. "Ferdinand Trendelenburg 60 Jahre." *Physikalische Blätter* 12:270.

Lunze, K., ed. 1981. *Heinrich Barkhausen. Barkhausen-Ehrung der Akademie der Wissenschaften der DDR und der Technischen Universität Dresden.* Dresden: Barkhausen-Komitee der Akademie der DDR.

MacLeod, Roy. 2000. "Sight and Sound on the Western Front: Surveyors, Scientists, and the 'Battlefield Laboratory.'" *War and Society* 18:23–46.

Maier, Helmut. 2007. *Forschung als Waffe: Rüstungsforschung in der Kaiser-Wilhelm-Gesellschaft und das Kaiser-Wilhelm-Institut für Metallforschung 1900–1945/48.* 2 vols. Göttingen: Wallstein Verlag.

Manegold, Karl-Heinz. 1970. *Universität, Technische Hochschule, Industrie.* Berlin: Duncker and Humblot.

Marinetti, Filippo Tommaso. 1909. "Le Futurisme." *Le Figaro* 20 (February): 1.

Markub, Abu. 1930. "Elektroakustikkens tidsalder." *Norsk radio* 8 (12): 259–263.

Mason, Joan. 2006. "Hertha Ayrton (1854–1923)." In Nina Byers and Gary Williams, eds., *Out of the Shadows: Contributions of Twentieth Century Women to Physics*, 15–25. Cambridge: Cambridge University Press.

Max Kohl AG. 1911. *Physical Apparatus, Price List No. 50.* Vols. 2 and 3. Chemitz: Max Kohl AG.

Maxwell, James Clerk. 1873. *A Treatise on Electricity and Magnetism.* 2 vols. Oxford: Clarendon Press.

Menges, Franz. 1987. "Löwenstein, Leo." *Neue Deutsche Biographie* 15: 106–107.

Meyenn, K. von , ed. 1994. *Quantenmechanik und Weimarer Republik.* Braunschweig, Wiesbaden: Vieweg and Sohn.

Meyer, Erwin. 1923. "Ponderomotorische Wirkungen von Tonwellen auf resonierende Membranen." *Annalen der Physik* 376 (16): 567–590.

Meyer, Erwin. 1931. "Die Klangspektren der Musikinstrumente." *Zeitschrift für technische Physik* 12: 606–611.

Meyer, Erwin. 1934. "Akustische Meßtechnik." In Erich Waetzmann, ed., *Handbuch der Experimentalphysik: Technische Akustik*, vol. 1, 73–159. Leipzig: Akademische Verlagsgesellschaft.

Meyer, Erwin. 1938. "Erich Waetzmann zum Gedächtnis." *Akustische Zeitschrift* 3 (5): 241–244.

Meyer, Erwin. 1939. *Electro-acoustics*. London: Bell and Sons.

Meyer, Erwin, and Erich Waetzmann. 1936. "Die Bedeutung der Akustik im Rahmen der gesamten Physik und Technik." *Akustische Zeitschrift* 1:114–118.

Michel, Andrée. 1994. "Richard Feldtkeller 1901–1981." In Ernst Feldtkeller and Herbert Goetzeler, eds., *Pioniere der Wissenschaft bei Siemens*, 108–112. Erlangen: Publicis MCD Verlag.

Mielert, Helmut. 1985. "Lichte, Hugo." *Neue Deutsche Biographie* 14:448–449.

Mühl-Benninghaus, Wolfgang. 1999. *Das Ringen um den Tonfilm—Strategie der Elektro- und der Filmindustrie in den 20er und 30er Jahren*. Düsseldorf: Droste Verlag.

The National Research Council Committee on Acoustics. 1922. "Certain Problems in Acoustics." *Bulletin of the National Research Council* 23, vol. 4, part 5, November.

Neumann, Hans, and Ferdinand Trendelenburg. 1931. "Über Hochleistungsblatthaller." *Zeitschrift für Hochfrequenztechnik* 37:149–151.

Otto, Rudolf. 1935. "Das Rauschen von Kohlemikrophonen." *Hochfrequenztechnik und Elektroakustik* 45:187–198.

Pantalony, David. 2009. *Altered Sensations: Rudolph Koenig's Acoustical Workshop in Nineteenth Century Paris*. Dordrecht: Springer.

Parey, Walter. 1935. "Technik ist Dienst am Volke! Rückblick auf die 73. Hauptversammlung des VDI mit dem 1. Tag der deutschen Technik und der 25-Jahrfeier der Technischen Hochschule Breslau." *Zeitschrift des Vereins Deutscher Ingenieure* 79 (27): 819–830.

Personalverzeichnis der Sächsischen Technischen Hochschule für das Wintersemester 1921/22. 1922. Dresden: Technische Hochschule Dresden.

Petersen, Sonja. 2011. *Vom "Schwachstarktastenkasten" und seinen Fabrikanten: Wissensräume im Klavierbau 1830 bis 1930*. Münster: Waxmann.

Phalkey, Jahnavi. 2013. "Focus: Science, History, and Modern India— Introduction." *Isis* 104 (2): 330–336.

Pickering, A., ed. 1992. *Science as Practice and Culture*. Chicago: University of Chicago Press.

Pinch, T., and K. Bijsterveld, eds. 2012. *The Oxford Handbook of Sound Studies*. New York: Oxford University Press.

Pinch, Trevor, and Frank Trocco. 2001. *Analog Days: The Invention and Impact of the Moog Synthesizer*. Cambridge, MA: Harvard University Press.

Poincaré, Henri. 1907. "Étude du récepteur téléphonique." *L'Éclairage électrique* 14:221–234, 257–262, 329–338, 365–372, 401–404.

Prakash, Gyan. 1999. *Another Reason: Science and the Imagination of Modern India*. Princeton, NJ: Princeton University Press.

Prandtl, Ludwig. 1921. "Bemerkungen über den Flugzeugschall." *Zeitschrift für technische Physik* 2 (9): 244–245.

Programm der Württembergischen Technischen Hochschule Stuttgart für das Studienjahr 1929/30. 1929. Stuttgart: Technische Hochschule Stuttgart.

Programm der Württembergischen Technischen Hochschule Stuttgart für das Studienjahr 1933/34. 1933. Stuttgart: Technische Hochschule Stuttgart.

Raman, Chandrasekhara Venkata. 1927. "Musikinstrumente und ihre Klänge." In Ferdinand Trendelenburg, ed., *Akustik*, vol. 8 of Handbuch der Physik, 354–424. Berlin: Julius Springer.

Ramaseshan, S., ed. 1988. *Acoustics*. Vol. 2. Scientific papers of CV Raman. Bangalore: Indian Academy of Sciences.

Rammer, Gerhard. 2004. "Die Nazifizierung und Entnazifizierung der Physik an der Universität Göttingen." PhD diss., University of Göttingen.

Ramsauer, Carl. 1943. "Die Schlüsselstellung der Physik für Naturwissenschaft, Technik und Rüstung." *Naturwissenschaften* 31 (25/26): 285–288.

Ramsauer, Carl. 1949. *Physik, Technik, Pädagogik: Erfahrungen und Erinnerungen*. Karlsruhe: Braun.

Ramsauer, Carl. 1965. "Physikalische Technik und Forschung–Einleitung." In Bruno Schweder, ed., *Forschen und schaffen. Beiträge der AEG zur Entwicklung der Elektrotechnik bis zum Wiederaufbau nach dem zweiten Weltkrieg*, vol. 3, 401–403. Berlin: AEG.

Rasch, Manfred. 2006. "Science and the Military: The Kaiser Wilhelm Foundation for Military-Technical Research." In Jeffrey Alan Johnson and Roy MacLeod, eds., *Frontline and Factory: Comparative Perspectives on the Chemical Industry at War, 1914–1924*, 179–202. Dordrecht: Springer.

Rasmussen, Anne. 2010. "Science and Technology." In John Horne, ed., *A Companion to World War I*, 307–322. Malden, MA: Wiley-Blackwell.

Ratcliffe, John Ashworth. 1971. "William Henry Eccles, 1875–1966." *Biographical Memoirs of Fellows of the Royal Society. Royal Society (Great Britain)* 17:195–214.

Rayleigh, Lord (John William Strutt). 1877. *The Theory of Sound*. Vol. 1. London: Macmillan.

Rayleigh, Lord (John William Strutt). 1878. *The Theory of Sound*. Vol. 2. London: Macmillan.

Rayleigh, Lord (John William Strutt). 1880. "Note on the Theory of the Induction Balance." *Report of the Annual Meeting*, vol. 50, 472–473. London: British Association for the Advancement of Science.

Rayleigh, Lord (John William Strutt). 1882. "A Telephone-Experiment." *Philosophical Magazine* 8:344.

Rayleigh, Lord (John William Strutt). 1894. *The Theory of Sound*. 2nd ed., vol. 1. London: Macmillan.

Rayleigh, Lord (John William Strutt). 1896. *The Theory of Sound*. 2nd ed., vol. 2. London: Macmillan.

Rehding, Alexander. 2000. "The Quest for the Origins of Music in Germany circa 1900." *Journal of the American Musicological Society* 52 (2): 345–385.

Reichardt, Walter. 1981. "Bewertung von Schall–von H. Barkhausen bis zur Gegenwart." In Klaus Lunze, ed., *Heinrich Barkhausen. Barkhausen-Ehrung der Akademie der Wissenschaften der DDR und der Technischen Universität Dresden*, 81–96. Dresden: Barkhausen-Komitee der Akademie der DDR.

Reichel, Hans. 1955. "Broemser, Philipp." *Neue Deutsche Biographie* 2: 630.

Reiher, Hermann. 1927. "Mittel zur Erzielung guter Raumakustik." *Gesundheits-Ingenieur* 50: 127.

Reiher, Hermann. 1932a. "Das Institut für Schall- und Wärmeforschung." *Bauwelt* 23 (13): 335–339.

Reiher, Hermann. 1932b. "Das Institut für Schall- und Wärmeforschung der Technischen Hochschule Stuttgart." *Zeitschrift des Vereins Deutscher Ingenieure* 76 (11): 277–278.

Reiher, Hermann. 1932c. *Über den Schallschutz durch Baukonstruktionsteile*. Beihefte zum Gesundheits-Ingenieur 2, vol. 11. Munich: Oldenbourg.

Reinhardt, Carsten. 1997. *Forschung in der chemischen Industrie: Die Entwicklung synthetischer Farbstoffe bei BASF und Hoechst, 1863 bis 1914*. Freiberg: TU Bergakademie.

Rheinberger, Hans-Jörg. 1997. *Towards a History of Epistemic Things—Synthesizing Proteins in the Test Tube*. Stanford, CA: Stanford University Press.

Richter, Steffen. 1972. *Forschungsförderung in Deutschland 1920–1936, dargestellt an Beispielen der Notgemeinschaft der Deutschen Wissenschaft und ihrem Wirken für das Fach Physik*. Düsseldorf: VDI-Verlag.

Riegger, Hans. 1912. "Über gekoppelte Kondensatorkreise bei sehr kurzer Funkenstrecke." *Jahrbuch der drahtlosen Telegraphie und Telephonie* 5:35–59.

Rihl, Wilhelm. 1911. "Über die Schallintensität des tönenden Lichtbogens." *Annalen der Physik* 341 (13): 647–680.

Rössler, Eberhard. 2006. *Die Sonaranlagen der deutschen Unterseeboote.* 2nd ed. Bonn: Bernhard and Graefe.

Rotta, Julius C. 1990. *Die Aerodynamische Versuchsanstalt in Göttingen, ein Werk Ludwig Prandtls: Ihre Geschichte von den Anfängen bis 1925.* Göttingen: Vandenhoeck and Ruprecht.

Ruhmer, Ernst. 1901. "Kinematographische Flammenbogenaufnahmen und das Photographophon, ein photographischer Phonograph." *Annalen der Physik* 310 (8): 803–810.

Rukop, Hans, and Jonathan Zenneck. 1914. "Der Lichtbogengenerator im Wechselstrombetrieb." *Annalen der Physik* 349 (9): 97–111.

Rupp, Hans. 1921a. "Grundsätzliches über Eignungsprüfungen." *Beihefte zur Zeitschrift für angewandte Psychologie* 29:32–62.

Rupp, Hans. 1921b. "Aus der Psychotechnik des subjektiven Schallmeßverfahrens." *Beihefte zur Zeitschrift für angewandte Psychologie* 29:131–149.

Russolo, Luigi. (1913) 1916. *L'arte dei Rumori: Edizione futuriste di poesía.* Milan: Corso Venezia.

Russolo, Luigi. (1913) 1986. *The Art of Noises.* Translated from the Italian with an introduction by Barclay Brown. New York: Pendragon Press.

Sabra, Abdelhamid I. 1996. "Situating Arabic Science: Locality versus Essence." *Isis* 87 (4): 654–670.

Salinger, Hans. 1930. "Das Heinrich-Hertz-Institut für Schwingungsforschung in Berlin." *Telegraphen- und Fernsprech-Technik* 7:216–218.

Salinger, Hans. 1934. "Fernübertragung von Schall. A. Fernübertragung auf Leitungen." In Erich Waetzmann, eds., *Handbuch der Experimentalphysik: Technische Akustik,* vol. 2, 257–270. Leipzig: Akademische Verlagsgesellschaft.

Schafer, R. Murray. 1994. *The Soundscape: Our Sonic Environment and the Tuning of the World.* Rochester, VT: Destiny Books.

Scharstein, Ernst. 1929. "Beiträge zur Raumakustik." *Annalen der Physik* 394 (2): 163–193.

Scharstein, Ernst, and Walter Schindelin. 1929. "Beiträge zur Raumakustik (Die Akustik der Aula der Albert-Ludwigs-Universität in Freiburg i Br.)." *Annalen der Physik* 394 (2): 194–200.

Schatzberg, Eric. 2006. "Technik Comes to America: Changing Meanings of Technology before 1930." *Technology and Culture* 47 (3): 487–512.

Scheel, Karl. 1918. "Die Tätigkeit der Physikalisch-Technischen Reichsanstalt im Jahre 1917." *Naturwissenschaften* 6 (37): 541–546.

Scheel, Karl. 1919. "Die Tätigkeit der Physikalisch-Technischen Reichsanstalt im Jahre 1918." *Naturwissenschaften* 7 (52): 997–1002.

Schenk, Dietmar. 1997. "Die Berliner Bundfunkversuchsstelle (1928–1935): Zur Geschichte und Rezeption einer Institution aus der Frühzeit von Rundfunk und Tonfilm." *Rundfunk und Geschichte* 23:124–126.

Schindelin, Walter. 1929. "Beiträge zur Raumakustik." *Annalen der Physik* 394 (2): 129–162.

Schirrmacher, Arne. 2009. "Von der Geschossbahn zum Atomorbital? Möglichkeiten der Mobilisierung von Kriegs- und Grundlagenforschung füreinander in Frankreich, Großbritannien und Deutschland, 1914–1924." In Matthias Berg, Jens Thiel, and Peter Th. Walther, eds., *Mit Feder und Schwert: Militär und Wissenschaft— Wissenschaftler und Krieg*, 155–175. Stuttgart: Franz Steiner.

Schirrmacher, Arne. 2014. "Die Physik im Großen Krieg." *Physik Journal* 13 (7): 43–48.

Schmaltz, Florian. 2013. "Militärische Problembündel." Unpublished manuscript, 18 pp.

Schmithals, Friedmann. 1980. "Carl Ramsauer und das Dritte Reich." *Physikalische Blätter* 36:345.

Schmucker, Georg. 2000. "Jonathan Zenneck 1871–1959. Eine technisch-wissenschaftliche Biographie." PhD diss., Universität Stuttgart.

Schoen, Lothar. 1994. "Karl Küpfmüller 1897–1977." In Ernst Feldtkeller and Herbert Goetzeler, eds., *Pioniere der Wissenschaft bei Siemens*, 96–102. Erlangen: Publicis MCD Verlag.

Schottky, Walter. 1918. "Über spontane Stromschwankungen in verschiedenen Elektrizitätsleitern." *Annalen der Physik* 362:541–567.

Schottky, Walter. 1924. "Vorführung eines neuen Lautsprechers. I." *Physikalische Zeitschrift* 25:672–675.

Schottky, Walter. 1926. "Das Gesetz des Tiefenempfangs in der Akustik und Elektroakustik." *Zeitschrift für Physik* 36:689–736.

Schottky, Walter. 1927. "Elektroakustik." In Karl Willy Wagner, ed., *Die wissenschaftlichen Grundlagen des Rundfunkempfangs*, 60–141. Berlin: Springer.

Schröder, Reinald. 1993. "Die 'schöne deutsche Physik' von Gustav Hertz und der 'weiße Jude' Heisenberg—Johannes Starks ideologischer Antisemitismus." In Helmut Albrecht, ed., *Naturwissenschaft und Technik in der Geschichte*, 327–341. Stuttgart: GNT-Verlag.

Schrödinger, Erwin. 1917. "Zur Akustik der Atmosphäre." *Physikalische Zeitschrift* 18:445–453, 567.

Schrödinger, Erwin. 1926. "Quantisierung als Eigenwertproblem (Vierte Mitteilung)." *Annalen der Physik* 386:109–139.

Schroeder, Manfred. 1972. "Erwin Meyer—1899–1972." *Journal of the Acoustical Society of America* 51:1489.

Schroeder, Manfred. 2000. "Erwin Meyer—Professor für Physik an der Universität Göttingen." *Gottinger Jahrbuch* 48:181–183.

Schubert, Georg Oskar. 1927. "Grundlegendes zu Untersuchungen an Mikrophonen." *Elektrische Nachrichten-Technik* 4:139–154.

Schubert, Helmut. 1987. "Mandelung, Erwin." *Neue Deutsche Biographie* 15: 628.

Schüller, Eduard. 1965. "Das Magnetophon." In Bruno Schweder, ed., *Forschen und schaffen. Beiträge der AEG zur Entwicklung der Elektrotechnik bis zum Wiederaufbau nach dem zweiten Weltkrieg*, vol. 3, 428–429. Berlin: AEG.

Schultrich, Helga. 1982. "Die Herausbildung des Industriephysikers im kapitalistischen Deutschland, dargestellt am Beispiel des Siemens- und des Zeiss-Konzerns." PhD diss., TU Dresden.

Schuster, Karl. 1934. "Grundbegriffe der technischen Akustik: Das Schallfeld." In Erich Waetzmann, ed., *Handbuch der Experimentalphysik: Technische Akustik*, vol. 1, 1–72. Leipzig: Akademische Verlagsgesellschaft.

Schwandt, Erich. 1934. "Lautsprecher-Laboratorium unter freiem Himmel." *Funk* 32:559–560.

Schweder, B., ed. 1965. *Forschen und schaffen. Beiträge der AEG zur Entwicklung der Elektrotechnik bis zum Wiederaufbau nach dem zweiten Weltkrieg*. 3 vols. Berlin: AEG.

Science Museum. 1935. *Noise Abatement Exhibition*, 31st May–30th June. London: Science Museum, South Kensington.

Serchinger, Reinhard W. 2008. *Walter Schottky: Atomtheoretiker und Elektrotechniker*. Diepholz: GNT-Verlag.

Seth, Suman. 2010. *Crafting the Quantum: Arnold Sommerfeld and the Practice of Theory, 1890–1926*. Cambridge, MA: MIT Press.

Shapin, Steven. 1998. "Placing the View from Nowhere: Historical and Sociological Problems in the Location of Science." *Transactions of the Institute of British Geographers* 23 (1): 5–12.

Shapin, Steven. 2004. "Who Is the Industrial Scientist? Commentary from Academic Sociology and from the Shop-Floor in the United States, ca. 1900–ca. 1970." In Karl Grandin and Nina Wormbs, eds., *The Science–Industry Nexus: History, Policy, Implications*, 337–363. Sagamore Beach, MA.: Science History Publications.

Shower, Edmund G., and Rulon Biddulph. 1931. "Differential Pitch Sensitivity of the Ear." *Journal of the Acoustical Society of America* 3 (2A): 275–287.

Simon, Hermann Theodor. 1898. "Akustische Erscheinungen am electrischen Flammenbogen." *Annalen der Physik* 300 (2): 233–239.

Simon, Hermann Theodor. 1906. "Zur Theorie des selbsttönenden Lichtbogens." *Physikalische Zeitschrift* 7 (13): 433–445.

Simon, Hermann Theodor. 1911. *Der elektrische Lichtbogen—Experimentalvortrag.* Leipzig: Hirzel.

Smith, Emily Mary, and Frederic Charles Bartlett. 1919. "On Listening of Sounds of Weak Intensity, Part I." *British Journal of Psychology* 10 (1): 101–129.

Smith, Emily Mary, and Frederic Charles Bartlett. 1920. "On Listening of Sounds of Weak Intensity, Part II." *British Journal of Psychology* 10 (2): 133–165.

Sound Ranging. 1917. Pamphlet issued by the General Staff, H.B.M. Government, March 1917, 9 pp. Available at http://www.defencesurveyors.org.uk/Images/Historical/WWI/Sound_Ranging_1917.pdf

Sprung, Helga, and Lothar Sprung. 2006. *Carl Stumpf—Eine Biographie: Von der Philosophie zur Experimentellen Psychologie.* Munich: Profil-Verlag.

Staley, Richard. 2005. "On the Co-Creation of Classical and Modern Physics." *Isis* 96 (4): 530–558.

Stange, Joachim. 1989. *Die Bedeutung der elektroakustischen Medien für die Musik im 20. Jahrhundert.* Pfaffenweiler: Centaurus.

Stangl, Burkhard. 2000. *Ethnologie im Ohr: Die Wirkungsgeschichte des Phonographen.* Vienna: WUV, Univ.-Verl.

Steinmetz, Charles Proteus, and Ernst J. Berg. 1897. *Theory and Calculation of Alternating Current Phenomena.* New York: W. J. Johnston.

Stenzel, Heinrich. 1922. "Über die Darstellbarkeit einer Matrix als Produkt von zwei symmetrischen Matrizen, als Produkt von zwei alternierenden Matrizen und als Produkt von einer symmetrischen und einer alternierenden Matrix." *Mathematische Zeitschrift* 15: 1–25.

Stenzel, Heinrich. 1926. "Akustische Lotmethoden." *Werft, Reederei, Hafen* 7:117–121, 139–143.

Sterne, Jonathan. 2003. *The Audible Past: Cultural Origins of Sound Reproduction.* Durham, NC: Duke University Press.

Sterne, J., ed. 2012. *The Sound Studies Reader.* London: Routledge.

Stewart, George W. 1919a. "Location of Aircraft by Sound." *Physical Review* 14 (2): 166–167.

Stewart, George W. 1919b. "Propagation of Sound in an Irregular Atmosphere." *Physical Review* 14 (3): 376–378.

Stewart, George W. 1921. "An Acoustic Wave Filter." *Physical Review* 17 (3): 382–384.

Stewart, George W. 1922. "Acoustic wave filters." *Physical Review* 20 (6): 528–551.

Stewart, George W. 1931. "Problems Suggested by an Uncertainty Principle in Acoustics." *Journal of the Acoustical Society of America* 2 (3): 325–329.

Stewart, John Q. 1922. "An Electrical Analogue of the Vocal Organs." *Nature* 110:311–312.

Strutt, Maximilian J. O. 1934. "Raumakustik." In Erich Waetzmann, ed., *Technische Akustik—Erster Teil (Handbuch der Experimentalphysik)*, 443–512. Leipzig: Akademische Verlagsges.

Stumpf, Carl. 1883. *Tonpsychologie.* Vol. 1. Leipzig: Hirzel.

Stumpf, Carl. 1890. *Tonpsychologie.* Vol. 2. Leipzig: Hirzel.

Stumpf, Carl. 1908. "Das Berliner Phonogrammarchiv." *Internationale Wochenschrift für Wissenschaft, Kunst und Technik* 2:225–246.

Stumpf, Carl. 1911. *Die Anfänge der Musik.* Leipzig: Johann Ambrosius Barth.

Stumpf, Carl. 1926. *Die Sprachlaute. Experimentell-phonetische Untersuchungen. Nebst einem Anhang über Instrumentalklänge.* Berlin: Julius Springer.

Stumpp, Hermann. 1936. *Experimentalbeitrag zur Raumakustik.* Beihefte zum Gesundheits-Ingenieur 2, vol. 17. Munich: Oldenbourg.

"Submarines Betrayed by Sound Waves: Detecting the Presence of Underwater Craft Fifty-five Miles Away by Microphone Devices." 1915. *Scientific American* 113 (16 October): 333, 346.

Szöllösi-Janze, Margit. 1998. *Fritz Haber 1868–1934: Eine Biographie.* Munich: Beck.

Tekniska fysikers förening. 1982. *Teknisk fysik i Sverige—Uppsatser tillägnade Professor Gudmund Borelius.* Stockholm: LiberTryck.

Thompson, Emily. 1997. "Dead Rooms and Live Wires: Harvard, Hollywood and the Deconstruction of Architectural Acoustics, 1900–1930." *Isis* 88 (4): 596–626.

Thompson, Emily. 2002. *The Soundscape of Modernity: Architectural Acoustics and the Culture of Listening in America, 1900–1933*. Cambridge, MA.: MIT Press.

Thompson, John B. 1995. *The Media and Modernity: A Social Theory of the Media*. Stanford, CA: Stanford University Press.

Tobies, Renate. 1997. *"Aller Männerkultur zum Trotz": Frauen in Mathematik und Naturwissenschaften*. Frankfurt am Main: Campus.

Trautwein, Friedrich. 1937. "Dynamische Probleme der Musik bei Feiern unter freiem Himmel." *Deutsche Musikkultur* 2 (1): 33–44.

Trendelenburg, Ferdinand. 1923. "Wirkungsweise und Anwendung des Thermophons." *Wissenschaftliche Veröffentlichungen aus dem Siemens-Konzern* 3 (1): 212–225.

Trendelenburg, Ferdinand. 1924. "Objektive Klangaufzeichnung mittels des Kondensatormikrophons." *Wissenschaftliche Veröffentlichungen aus dem Siemens-Konzern* 3 (2): 43–66.

Trendelenburg, Ferdinand. 1925. "Über eine Methode zur objektiven Lautsprecheruntersuchung." *Wissenschaftliche Veröffentlichungen aus dem Siemens-Konzern* 4 (2): 200–204.

Trendelenburg, Ferdinand. 1926. "Beiträge zu Schallfeldmessungen." *Wissenschaftliche Veröffentlichungen aus dem Siemens-Konzern* 5 (2): 120–134.

Trendelenburg, F., ed. 1927. *Akustik*, vol. 8 of Handbuch der Physik. Berlin: Julius Springer.

Trendelenburg, Ferdinand. 1928. "Über physikalische Eigenschaften der Herztöne." *Wissenschaftliche Veröffentlichungen aus dem Siemens-Konzern* 6 (2): 184–208.

Trendelenburg, Ferdinand. 1931. "Objektive Messung und subjektive Beobachtung von Schallvorgängen." *Naturwissenschaften* 19 (47): 937–940.

Trendelenburg, Ferdinand. 1932. *Die Fortschritte der physikalischen und technischen Akustik*. Leipzig: Akad. Verlagsges.

Trendelenburg, Ferdinand. 1934. *Die Fortschritte der physikalischen und technischen Akustik*. 2nd enlarged ed. Leipzig: Akad. Verlagsges.

Trendelenburg, Ferdinand. 1935. *Klänge und Geräusche—Methoden und Ergebnisse der Klangforschung*. Berlin: Julius Springer.

Trendelenburg, Ferdinand. 1939. *Einführung in die Akustik*. Berlin: Julius Springer.

Trendelenburg, Ferdinand. 1950. "Hermann Backhaus zum 65. Geburtstag." *Physikalische Blätter* 6:417–418.

Trendelenburg, Ferdinand. 1951. "Hans Gerdien †." *Physikalische Blätter* 7 (5): 221–222.

Trendelenburg, Ferdinand. 1966. "Zwischen Tokyo und Chicago: Meist heitere Erinnerungen aus dem Leben eines Physikers für Kinder, Enkel und Freunde niedergeschrieben." Erlangen: Unpublished autobiography.

Trendelenburg, Ferdinand. 1975. *Aus der Geschichte der Forschung im Hause Siemens. Technikgeschichte in Einzeldarstellungen.* Düsseldorf: VDI-Verlag.

Trischler, Helmuth. 1996. "Die neue Räumlichkeit des Krieges: Wissenschaft und Technik im 1. Weltkrieg." *Berichte zur Wissenschaftsgeschichte* 19:95–103.

Tröger, Joachim. 1930. "Die Schallaufnahme durch das äußere Ohr." *Physikalische Zeitschrift* 31:26–47.

Trumpy, Bjørn. 1930. *Akustikk—utvalgte forelesninger for elektroavdelingens, 4. årskurs (linje for svakstrøm) 1. del.* Trondheim: Tapir forlag.

Tucker, William S., and E. Tabor Paris. 1921. "A Selective Hot-Wire Microphone." *Philosophical Transactions of the Royal Society of London.* Series A, Mathematical and Physical Sciences 221:389–430.

Ungern-Sternberg, Jürgen von, and Wolfgang von Ungern-Sternberg. 1996. *Der Aufruf "An die Kulturwelt!" das Manifest der 93 und die Anfänge der Kriegspropaganda im Ersten Weltkrieg.* Stuttgart: Franz Steiner.

Van der Kloot, William. 2005. "Lawrence Bragg's Role in the Development of Sound Ranging." *Notes and Record of the Royal Society* 59:273–284.

Van der Kloot, William. 2011. "Mirrors and Smoke: A. V. Hill, His Brigands, and the Science of Anti-aircraft Gunnery in World War I." *Notes and Record of the Royal Society* 65:393–410.

Verleihung der Ehrendoktorwürde der Technischen Universität Berlin an Herrn Professor Erwin Meyer. 1958. Berlin: Rektor und Senat der Technischen Universität Berlin, November 18.

Vierling, Oskar. 1936. *Das elektroakustische Klavier.* Berlin: VDI-Verlag.

Vierling, Oskar. 1938a. *Eine neue elektrische Orgel.* Berlin: Technische Hochschule Habilitationsschrift.

Vierling, Oskar. 1938b. "Erfahrungen mit einer 5 kW Großlautsprecheranlage auf der Burg zu Nürnberg." *Akustische Zeitschrift* 3 (2): 93–96.

Vieweg, Richard. 1948. "Technische Physik." *Physikalische Blätter* 4:16–20.

Voigt, Wolfgang. 1988. "Oskar Vierling, ein Wegbereiter der Elektroakustik für den Musikinstrumentenbau." Pts. 1 and 2. *Das Musikinstrument* 37 (1/2): 214–221; (3/4): 172–176.

Waetzmann, Erich. 1912. *Die Resonanztheorie des Hörens als Beitrag zur Lehre von den Tonempfindungen*. Braunschweig: Vieweg and Sohn.

Waetzmann, Erich. 1921a. "Die Entstehung und Art des Flugzeugschalles." *Zeitschrift für technische Physik* 2 (6): 166–172.

Waetzmann, Erich. 1921b. "Das Abhören von Flugzeugschall." *Zeitschrift für technische Physik* 2 (7): 191–194.

Waetzmann, Erich. 1927. "Zur Ausbreitung elastischer Wellen in der Erdoberfläche." *Naturwissenschaften* 15:401–403.

Waetzmann, E., ed. 1934a. *Handbuch der Experimentalphysik: Technische Akustik*. 2 vols. Leipzig: Akademische Verlagsgesellschaft.

Waetzmann, Erich. 1934b. *Schule des Horchens*. Leipzig: B. G. Teubner.

Wagner, Karl Willy. 1910. "Der Lichtbogen als Wechselstromerzeuger." PhD diss., University of Göttingen.

Wagner, Karl Willy. 1912. "Über die Verbesserung des Telephons." *Jahrbuch der drahtlosen Telegraphie und Telephonie* 1 (1): 38–43.

Wagner, Karl Willy. 1919. "Spulen- und Kondensatorleitungen." *Archiv für Elektrotechnik* 8 (2/3): 61–92.

Wagner, K. W., ed. 1927. *Die wissenschaftlichen Grundlagen des Rundfunkempfangs*. Berlin: Springer.

Wagner, Karl Willy. 1930. "Das Heinrich-Hertz-Institut für Schwingungsforschung." *Elektrische Nachrichten-Technik* 7:174–191.

Wagner, Karl Willy. 1931. "Geräusch und Lärm." *Sitzungsberichte der Preussischen Akademie der Wissenschaften, Physikalisch-Mathematische Klasse* 9:154–165.

Wagner, Karl Willy. 1936a. "Grundlagen der Lärmabwehr." *Forschungen und Fortschritte* 12 (1): 12–13.

Wagner, Karl Willy. 1936b. *Ein neues elektrisches Sprechgerät zur Nachbildung der menschlichen Vokale*. Abhandlungen der Preußischen Akademie der Wissenschaften, Physikalisch-Mathematische Klasse 2. Berlin: Akademie der Wissenschaften.

Wagner, Karl Willy. 1937. "Max Wien zum 70. Geburtstag." *Naturwissenschaften* 25 (5): 65–67.

Warwick, Andrew. 2003. *Masters of Theory: Cambridge and the Rise of Mathematical Physics*. Chicago: University of Chicago Press.

Wegel, Raymond Lester. 1930. "Theory of Vibration of the Larynx." *Bell System Technical Journal* 9 (1): 207–227.

Wegel, Raymond Lester, and Clarence Edward Lane. 1924. "The Auditory Masking of One Pure Tone by Another and Its Probable Relation to the Dynamics of the Inner Ear." *Physical Review* 23:266–285.

Wein, Adelheid. 2011. "Heinrich Barkhausen und die Anfänge der wissenschaftlichen Schwachstromtechnik." Magisterarbeit in der Philosophischen Fakultät I, University of Regensburg.

Welker, Heinrich. 1974. "Ferdinand Trendelenburg," 229–233. Munich: *Jahrbuch der Bayerischen Akademie der Wissenschaften.*

Wigge, Heinrich. 1934. *Technisches Hilfsbuch für Gemeinschaftsempfang, Hörberatung und Funkschutz.* Stuttgart: Franckh.

Wien, Max. 1889. "Über die Messung der Tonstärke." *Annalen der Physik* 272 (4): 834–857.

Wien, Max. 1891a. "Das Telephon als optischer Apparat zur Strommessung." *Annalen der Physik* 278 (4): 593–621.

Wien, Max. 1891b. "Das Telephon als optischer Apparat zur Strommessung. II." *Annalen der Physik* 280 (12): 681–688.

Wien, Max. 1891c. "Messung der Inductionsconstanten mit dem „optischen Telephon." *Annalen der Physik* 280 (12): 689–712.

Wien, Max. 1903. "Ueber die Empfindlichkeit des menschlichen Ohres für Töne verschiedener Höhe." *Pflügers Archiv für Physiologie* 97:1–57.

Wijfjes, Huub. 2014. "Spellbinding and Crooning: Sound Amplification, Radio and Political Rhetoric in International Comparative Perspective, 1900–1945." *Technology and Culture* 55 (1): 148–185.

Wille, Peter C. 2005. *Sound Images of the Ocean in Research and Monitoring.* Berlin: Springer.

Winterbotham, Harold St John Loyd. 1918. *Survey on the Western Front.* Preliminary Report, 20 December. Directorate of Military Survey, Tech Library No. 536:355.48 "1914–1918." Maps General Headquarters. http://www.defencesurveyors.org.uk/Historical/WWI/WWI.htm.

Withers, Charles W. J. 2001. *Geography, Science and National Identity: Scotland since 1520.* Cambridge: Cambridge University Press.

Wittje, Roland. 2000. "Experimentelle Tätigkeit und theoretische Konzepte: Heinrich Hertz zur elektrodynamischen Wirkung von Isolatoren." In Christoph Meinel, ed., *Instrument—Experiment: Historische Studien,* 180–191. Berlin: GNT-Verlag.

Wittje, Roland. 2003. "Acoustics, Atoms Smashing and Amateur Radio: Physics and Instrumentation at the Norwegian Institute of Technology in the Interwar Period." PhD diss., NTNU, Trondheim.

Wittje, Roland. 2013. "The Electrical Imagination: Sound Analogies, Equivalent Circuits, and the Rise of Electroacoustics, 1863–1939." *Osiris* 28:40–63.

Wittje, Roland. 2016. "Concepts and Significance of Noise in Acoustics: Before and after the Great War." *Perspectives on Science* 24 (1): 7–28.

Wolff, Stefan L. 2006. "Zur Situation der deutschen Universitätsphysik während des Ersten Weltkrieges." In Trude Maurer, ed., *Kollegen—Kommilitonen—Kämpfer: Europäische Universitäten im Ersten Weltkrieg*, 267–281. Stuttgart: Franz Steiner.

Wulff, Hans Jürgen, and Olaf Schumacher. 2001 "Warner, Fox und die Anfänge des Tonfilms." In Joachim-Felix Leonhardt, Hans-Werner Ludwig, Dietrich Schwarze, and Erich Straßner, eds., *Medienwissenschaft: ein Handbuch zur Entwicklung der Medien und Kommunikationsformen*, vol. 2, 1197–1208. Berlin: de Gruyter.

Yeang, Chen-Pang. 2012. "From Mechanical Objectivity to Instrumentalizing Theory: Inventing Radio Ionospheric Sounders." *Historical Studies in the Natural Sciences* 42 (3): 190–234.

Yeang, Chen-Pang. 2013. *Probing the Sky with Radio Waves: From Wireless Technology to the Development of Atmospheric Science*. Chicago: University of Chicago Press.

Zenneck, Jonathan. 1898. "Ein Versuch mit kreisförmigen Klangplatten." *Annalen der Physik* 302 (9): 170–176.

Zenneck, Jonathan. 1899a. "Ueber die freien Schwingungen nur annähernd vollkommener kreisförmiger Platten." *Annalen der Physik* 303 (1): 165–184.

Zenneck, Jonathan. 1899b. "Eine Methode zur Demonstration und Photographie von Stromcurven." *Annalen der Physik* 305 (12): 838–853.

Zenneck, Jonathan. 1904. "Objektive Darstellung von Stromkurven mit der Braunschen Röhre." *Annalen der Physik* 318 (4): 819–821.

Zenneck, Jonathan. 1905. *Elektromagnetische Schwingungen und drahtlose Telegraphie*. Stuttgart: Ferdinand Enke.

Zenneck, Jonathan. 1914. "Die Entstehung der Schwingungen bei der Lichtbogenmethode." *Annalen der Physik* 348 (4): 481–524.

Zenneck, Jonathan. 1925. "Die Lautsprecher bei der Eröffnung des Deutschen Museums." *Zeitschrift für Hochfrequenztechnik* 26: 177–179.

Zenneck, Jonathan. 1939. "W. Hahnemann." *Hochfrequenztechnik und Elektroakustik* 53 (6): 214.

Zenneck, Jonathan. 1960. *Erinnerungen eines Physikers*. Munich: privately printed.

Ziehm, Günther H.   1988. *Kiel—Ein frühes Zentrum des Wasserschalls*. Deutsche Hydrographische Zeitschrift, Ergänzungsheft Reihe B, Nr. 29. Hamburg: Deutsches Hydrographisches Institut.

Zimmerer, Jürgen, and Joachim Zeller. 2004. *Völkermord in Deutsch-Südwestafrika: der Kolonialkrieg (1904–1908) in Namibia und seine Folgen*. Berlin: Ch. Links.

Zimmerman, David. 1997. "Tucker's Acoustical Mirrors: Aircraft Detection before Radar." *War and Society* 15:73–99.

Zimmerman, David. 2001. *Britain's Shield: Radar and the Defeat of the Luftwaffe*. Stroud: Sutton.

Name Index

Subject Index

Radio, 8–9, 22–24, 59, 62, 65–66, 131,
    137–141, 146, 152, 156, 158–159,
    161, 172–176, 178–180, 190–191,
    198, 209
  tube, 18, 23, 71, 88, 106, 124–125,
    135, 140, 142–144, 150, 159, 203 (*see
    also* Valve [thermionic])
Radio Corporation of America (RCA),
    141
Radioelektrische Versuchsanstalt für
    Heer und Marine (Army and Navy
    Radioelectric Testing Laboratory), 65,
    91, 106
Raman scattering, 15
Ramsauer-Townsend effect, 192
Rauschen (electrical noise), 17–18, 145,
    203
Reichsgemeinschaft der technisch-
    wissenschaftlichen Arbeit (RTA), 176
Reichsmarineamt, 65
Reichsministerium für Volksaufklärung
    und Propaganda, 174, 177, 179
Reichspost, 57, 138–139, 142, 148, 150,
    159–160
Reichspostzentralamt, 160, 185
Reichs-Rundfunk-Gesellschaft (German
    Broadcasting Corporation), 138–139,
    160, 177
Reichssportfeld stadium, 182
Reichswaffenamt, 160, 185
Reichswehr, 113
Reichswoche ohne Lärm, 176
Relativity, theory of, 2, 26, 37, 144, 173,
    176, 192, 194–195
Ribbon loudspeaker, 145–147
Ribbon microphone, 145–147
Royal Air Force, 86
Royal Navy's Anti-Submarine Division,
    99
Rundfunkversuchsstelle, 158, 163, 183

Schallantenne (sound antenna), 107
Schallempfänger (sound receiver), 107

Schallfeld (sound field), 107–108, 159,
    167, 187, 192
Schallsender (sound transmitter),
    107–108
Schrotrauschen (shot noise), 144
Schwachstrom Laboratory, Dresden, 199
Schwachstromtechnik (low-current
    electrical engineering), 123–124
Schwingungsforschung (vibration and
    oscillation research), 24, 64, 124,
    157, 195, 199
Schwingungslehre (oscillation studies),
    160
Schwingungstechnik (oscillation
    technology), 26
Sendeleistung (transmitted power
    output), 108
Siemens and Halske (company), 16, 24,
    29, 57, 63, 81, 84, 92–93, 104, 106,
    113, 115, 118, 126, 135–138,
    140–150, 153–156, 158, 180,
    185–187, 193, 203, 206
Siemens loop oscillograph, 81–82, 86,
    132
Siemens Nernst Bechstein piano
    (Neo-Bechstein), 149, 163
Siemens Research Laboratories, 93, 115,
    134, 149, 187, 199, 209
Siemens Schuckertwerke, 142, 158
Siemensstadt district, Berlin, 135, 142
Signal Gesellschaft, 92–94, 107–108,
    114, 135, 140–141, 143, 151, 185
Signal School, 99
Singing arc, 58, 66, 203. *See also*
    Speaking arc
Society for the Suppression of
    Unnecessary Noise, 110
Sommerfeld school, 26
Sonar, 26, 98. *See also* Asdic
Sound-absorbing materials, 10, 125,
    132–133, 166
Sound locator, 75–77, 81, 83–85, 95,
    210

Transformations: Studies in the History of Science and Technology

Jed Z. Buchwald, general editor

William R. Newman and Anthony Grafton, editors, *Secrets of Nature: Astrology and Alchemy in Early Modern Europe*

Naomi Oreskes and John Krige, editors, *Science and Technology in the Global Cold War*

Gianna Pomata and Nancy G. Siraisi, editors, *Historia: Empiricism and Erudition in Early Modern Europe*

Alan J. Rocke, *Nationalizing Science: Adolphe Wurtz and the Battle for French Chemistry*

George Saliba, *Islamic Science and the Making of the European Renaissance*

Suman Seth, *Crafting the Quantum: Arnold Sommerfeld and the Practice of Theory, 1890–1926.*

William Thomas, *Rational Action: The Sciences of Policy in Britain and America, 1940–1960*

Leslie Tomory, *Progressive Enlightenment: The Origins of the Gaslight Industry 1780–1820*

Nicolás Wey Gómez, *The Tropics of Empire: Why Columbus Sailed South to the Indies*

Roland Wittje, *The Age of Electroacoustics: Transforming Science and Sound*